RISK ASSESSMENT
OF POWER SYSTEMS

Other Books in the IEEE Press Series on Power Engineering

Power System Protection
P. M. Anderson
1999   Hardcover   1344 pp   0-7803-3427-2

Understanding Power Quality Problems: Voltage Sags and Interruptions
Math H. J. Bollen
2000   Hardcover   576 pp   0-7803-4713-7

Electric Power Applications of Fuzzy Systems
Edited by M. E. El-Hawary
1998   Hardcover   384 pp   0-7803-1197-3

Principles of Electric Machines with Power Electronic Applications,
Second Edition
M. E. El-Hawary
2002   Hardcover   496 pp   0-471-20812-4

Analysis of Electric Machinery and Drive Systems, Second Edition
Paul C. Krause, Oleg Wasynczuk, and Scott D. Sudhoff
2002   Hardcover   643 pp   0-471-14326-X

Electric Power Systems: Analysis and Control
Fabio Saccomanno
2003   Hardcover   728 pp   0-471-23439-7

Electrical Insulation for Rotating Machines: Design, Evaluation, Aging, Testing, and Repair
G. C. Stone, E. A. Boulter, I. Culbert, H. Dhirani

RISK ASSESSMENT OF POWER SYSTEMS
Models, Methods, and Applications

WENYUAN LI, Ph.D., IEEE Fellow
British Columbia Transmission Corporation, Canada
Advisory Professor, Chongqing University, China

IEEE Press Series on Power Engineering
Mohamed E. El-Hawary, *Series Editor*

IEEE PRESS

A JOHN WILEY & SONS, INC., PUBLICATION

For general information on our other products and services please contact our Customer Care Department within the U.S. at 877-762-2974, outside the U.S. at 317-572-3993 or fax 317-572-4002.

Wiley also publishes its books in a variety of electronic formats. Some content that appears in print, however, may not be available in electronic format.

Library of Congress Cataloging-in-Publication Data is available.

ISBN 0-471-63168-X

10 9 8 7 6 5 4 3 2 1

To
Jun and my family

CONTENTS

PREFACE

The application of power system risk assessment has drawn ever-increasing interest in the electric utility industry, particularly since massive power outage events happened across the world in recent years. In essence, power systems behave probabilistically and system risk cannot be avoided fully. However, system risk can be evaluated and managed to an acceptable level in planning, design, operation, and maintenance activities.

This book originated from my deep involvement in this area, including development of models and methods and their actual applications. My previous employer, BC Hydro, and current company, British Columbia Transmission Corporation, have always provided me an encouraging atmosphere to carry on many projects for more than 10 years. Practical engineering problems and the deregulation process in the power market have created new challenges in system risk analysis. This enabled me to produce a considerable number of technical reports and papers that formed the foundation of the book. The experience from my academic career to the industrial environment made me recognize that a gap exists between power system risk theory and actual applications in utilities. The intent of this book is to fill that gap. The difficulties faced by practicing engineers include how to determine the component outage models that really reflect reality, how to deal with the uncertainty in statistical data, how to select the appropriate risk evaluation method for a specific case and, particularly, how to apply fundamentals and methodologies to each individual engineering issue.

Graduates who lack an understanding of real utility systems often find themselves at a loss regarding an industrial subject, even if they may have good knowledge in the risk theory and basic mathematics. This book has been structured to respond to these concerns. The primary emphasis is placed on the various application topics, although some new models and concepts have been also presented. The majority of examples that demonstrate the applications are based on actual projects that have been performed. This should be the most distinguishing feature of the book. It is my belief that the book will serve both engineers and advanced students who are or will be dedicated to power system risk assessment and management.

The book would not have been possible without my close association with the many friends, colleagues, professors, and engineers in the IEEE and the entire power field. My special appreciation goes to Roy Billinton, Yakout Mansour, Ebrahim Vaahedi, and Jerry Korczynski. The papers I coauthored with them are a part of the materials used in the book. Moreover, they have kindly offered me constant support and encouragement in my daily work. My sincere acknowledgment is extended to the following people for their support and help: Gerry Garnett, Don Gillespie, Steven Pai, Narayan Rau, Murty Bhavaraju, Lalit Goel, Bin Shi, Jiaqi Zhou, and Lichun Shu. I would also thank all the individuals whose publications are recorded in the References at the end of the book.

I am also grateful for the cooperation and assistance received from the IEEE Press and John Wiley & Sons, Inc., especially Anthony VenGraitis and Lisa Van Horn.

Finally, but never least, I would like to express my deepest gratitude to my wife, Jun, for her sacrifices and patience. She successfully arranged our lives in such a way that I could comfortably work nights and weekends

WENYUAN LI

Vancouver, Canada
August 2004

CHAPTER 1

INTRODUCTION

1.1 RISK IN POWER SYSTEMS

There is considerable overlap in the words "risk" and "reliability." In this book, it is
assumed that the two words have the identical implication. They are the two facets
of the same fact. Higher risk means lower reliability, and vice versa. Risk management has a wide-ranging content. The intent of the book is to discuss the models,
methods, and applications of risk assessment in physical power systems. Risks associated with business, finance, and life safety are not included in the discussion.

The probabilistic behavior of power systems is the root origin of risk. Random
failures of system equipment are generally outside the control of power system personnel. Loads always have uncertainties and it is impossible to obtain an exact load
forecast. Energy exports and imports under the deregulation environment depend
on the volatile power market. Consequences of power failures range from electricity interruptions in local areas to a possible widespread blackout. Economic impacts
due to supply interruptions are not restricted to loss of revenue by the utility or loss
of energy utilization by the customer but include indirect costs imposed on society
and the environment. Risk assessment has become a challenge and an essential
commitment in the power utility industry today.

Risk management includes at least the following three tasks:

1. Performing quantitative risk evaluation
2. Determining measures to reduce risk
3. Justifying an acceptable risk level

The purpose of quantitative evaluation is to create the indices representing system risk. A dictionary definition of risk is "the probability of loss or damage to hu-

Risk Assessment of Power Systems. By Wenyuan Li
ISBN 0-471-63168-X © 2005 the Institute of Electrical and Electronics Engineers, Inc.

man beings or assets." This definition can be used in general cases. However, a comprehensive risk index should not contain only the probability but a combination of probability and consequence. In other words, the risk evaluation of power systems should recognize not only the likelihood of failure events but also the severity and degree of their consequences. Utilities have dealt with system risks for a long time. The criteria and methods first used in practical applications were all deterministically based, such as the percentage reserve in generation capacity planning and the single-contingency principle in transmission planning. The deterministic criteria have served the power industry for years. The basic weakness is that they do not respond to the probabilistic nature of power system behavior, load variation, and component failures.

A measure to reduce system risk is generally associated with enhancement of the system. In order to determine a rational measure, both the impact of the measure on risk reduction and the cost needed to implement it should be quantified. A probabilistic economic analysis is usually required. In risk management, an important concept to be appreciated is that zero risk can never be reached since random failure events are uncontrollable. In many cases, a decision has to be made to accept a risk as long as it can be technically and financially justified. Selecting a rational measure to reduce risks or accepting a risk level is a decision-making process. It should be recognized that on the one hand, quantitative risk evaluation is the basis of this process, and on the other hand, the process is more than risk evaluation and requires technical, economic, societal, and environmental assessments.

The risk assessment of power systems can be applied to all the areas in electric power utilities, including:

- Quantified reliability evaluation in generation, transmission, and distribution systems
- Probabilistic criteria in system planning and operation
- Compromise between the system risk and the economic benefit in a decision-making process
- Equipment aging failure management
- Spare equipment strategy
- Reliability-centered maintenance
- Load-side risk management
- Performance-based rate policy
- Operation-risk monitoring
- Interruption damage cost assessment

Risk management and quantified risk assessment have become ever-increasingly important since the power industry entered the deregulation era. The new competition environment forces utilities to plan and operate their systems closer to the limit. The stressed operation conditions have led to deterioration in system reliability. In fact, a lot of power-outage events have occurred across the world in the past years. According to an EPRI (Electric Power Research Institute) report based on the national survey in all business sectors, the U.S. economy alone is losing between $104 and $164 billion a year due to power system outages [83]. Severe power outage events have happened frequently in recent years. For instance, a major system disturbance separated the Western Electricity Coordinating Council (WECC) system in the west of North America into four islands on August 10, 1996, interrupting electricity service to 7.5 million customers for a period of up to nine hours. The 1998 blackout at the Auckland central business district in New Zealand impacted 30 square blocks of the downtown area for about two months, resulting in lawsuits totaling $600 million against the utility. On August 14, 2003, the massive blackout in the east of North America covered eight states in the United States and two provinces in Canada, bringing about 50 million people into darkness for periods ranging from one to several days. These severe power outages let us realize that the single-contingency criterion (the N-1 principle) that has been used for many years in the power industry may not be sufficient to preserve a reasonable system reliability level. However, it is also commonly recognized that no utility can financially justify the N-2 or N-3 principle in power system planning. Obviously, one alternative is to bring risk management into the practice in planning, design, operation, and maintenance, keeping system risk within an acceptable range. On the other hand, the customers of the power industry have become more and more knowledgable about electric power systems. They understand that it is impossible to expect 100% continuity in the power supply without any risk of outages. However, they have the right to know the risk level, including information on how often, for how long, and how severely a power interruption event can happen to them on the average. To answer this question is one of the objectives of power system risk assessment.

1.2 BASIC CONCEPTS OF POWER SYSTEM RISK ASSESSMENT

1.2.1 System Risk Evaluation

Power system risk evaluation is generally associated with the following four tasks:

1. Determining component outage models
2. Selecting system states and calculating their probabilities
3. Evaluating the consequences of selected system states
4. Calculating the risk indices

A power system consists of many components, including generators, transmission lines, cables, transformers, breakers, switches, and a variety of reactive power source equipment. Component outages are the root cause of a system failure state. The first task in system risk evaluation is to determine component outage models. Component failures are classed into two categories: independent and dependent outages. Each category can be further classified according to the outage modes. In most cases, only repairable forced outages are considered, whereas in some cases, planned outages are also modeled. Aging failures have not been incorporated into the traditional risk evaluation. This book presents a modeling approach to aging failures and demonstrates examples of its application.

The second task is to select system failure states and calculate their probabilities. There are two basic methods for selecting a system state: state enumeration and Monte Carlo simulation. Both methods have merits and demerits. In general, if complex operating conditions are not considered and/or the failure probabilities of components are quite small, the state enumeration techniques are more efficient. When complex operating conditions are involved and/or the number of severe events is relatively large, Monte Carlo methods are often preferable.

The third task is to perform the analysis for system failure states and assess their consequences. Depending on the system under study, the analysis could be associated with the simple power balance, or the connectivity identification of a network configuration, or the complex calculation process, including the power flow, optimal power flow, or even transient and voltage-stability evaluation.

As mentioned earlier, risk is a combination of probability and consequence. With the information obtained in the second and third tasks, an index that truly represents system risk can be created. There are many possible risk indices for different purposes. Most of them are basically the expected value of a random variable, although a probability distribution can be calculated in some cases. It is important to appreciate that the expected value is not a deterministic parameter. It is the long-run average of the phenomenon under study. The expected indices serve as the risk indicators that reflect various factors, including component capacities and outages, load profiles and forecast uncertainties, system configurations and operational conditions, and so on.

According to system state analysis, power system risk assessment can be divided into two basic aspects: system adequacy and system security. Adequacy relates to the existence of sufficient facilities within the system to satisfy consumer load demand and system operational constraints. Adequacy is therefore associated with the static conditions that do not include system dynamic and transient processes. Security relates to the ability of the system to respond to dynamic and transient disturbances arising within the system. Security is therefore associated with the response of the system to whatever perturbations it is subject to. Normally, security evaluation requires the analysis of dynamic, transient, or voltage stability in the system. It should be pointed out that most of the risk evaluation techniques that have been used in the actual applications of utilities are in the domain of adequacy assessment. Some ideas for security assessment have been addressed recently. However, the practical application in this area is limited. Another fact is that most of the risk indices used in risk evaluation are inadequacy indices, not overall risk indices. The system indices that are based on historical outage statistics encompass the effect of both inadequacy and insecurity. It is important to recognize this fundamental difference in actual engineering applications.

A power system includes the three fundamental functions of generation, transmission (including substation), and distribution. Traditionally, the three functional zones are included in one utility. As reform in the power industry proceeds, the three functional zones have been gradually separated to form organizationally independent generation, transmission, and distribution companies in many countries. In either case, risk assessment can be, and is, conducted in each of these functional zones. The risk evaluation for an overall system, including generation, transmission, and distribution, is impractical because such a system is too enormous to handle in terms of the existing computing capacity and accuracy requirements. On the one hand, the calculation modeling and algorithms are quite different for the risk evaluation of a generation, transmission, substation, distribution system. On the other hand, many techniques have been successfully developed to perform the risk evaluation for composite generation and transmission systems or composite transmission and substation systems. In the case of a large-scale transmission system, it is reasonable to limit the study to an area or subsystem. Doing so can provide more realistic results than evaluating the whole system. This is due to the fact that a change or reinforcement in the network may considerably affect a local area but have little impact on remote parts of the system. The contribution to the overall reliability of a large system due to a local line addition or reconfiguration may be so small that it is masked by computational errors and, consequently, cannot be reflected in the risk change of the whole system. This contribution, however, can be a relatively large portion of the risk change in the local area.

Generally, it is necessary to assess the relative benefits of different alternatives, including the option of doing nothing. The level of analysis need not be any more complex than that which enables the relative merits to be assessed. The ability to include a high degree of precision in calculations should never override the inherent uncertainty in the data. An absolute risk index, although an ideal objective, is virtually impossible to evaluate. This does not weaken the necessity to objectively assess the relative merits of alternative schemes. This is an important point to be appreciated in power system risk evaluation.

1.2.2   Data in Risk Evaluation

The reliability data required in power system risk evaluation are the parameters of component outage models. They are basically calculated from historical statistics, although an engineering judgment based on individual equipment assessments is also used in some special cases. Collecting suitable data is at least as essential as developing risk evaluation methods.

The data requirements should reflect the need for risk assessments. The data must be sufficiently comprehensive to ensure that an evaluation method can be applied, but restrictive enough to ensure that unnecessary data are not collected. For the simple models, the data relates to the two main processes of component behavior, namely, the failure process and the restoration process. For more complex models, the data are associated with the transition rates between various states.

The quality of data is an important factor to consider in data collection. The usual saying of "garbage in and garbage out" refers to the fact that if the quality of data cannot be guaranteed, the results of risk evaluation will not make any sense. Outage statistics constitute a huge data pool and some bad or invalid records cannot be fully avoided in any database. Data processing is necessary to filter out bad data. A parameter estimation procedure is needed to acquire the input data of risk evaluation from raw statistics. This requires the suitable design of statistical data modeling.

Another characteristic of reliability data is its dynamic feature. The volume of outage records will increase with the time and, therefore, the average failure frequency and repair time for a piece of equipment or an equipment group will change from year to year. The reliability database should have a means of continuous updating and should be also flexible enough to output reports in a variety of formats. Figure 1.1 is an example of reliability data collection system. The abbreviation SETR in the figure stands for system equipment trouble report.

1.2.3   Unit Interruption Cost [71]

The basic function of a modern electric power system is to provide electric energy to its customers at the lowest possible cost and at an acceptable risk level. There is a conflict between economy and risk. Risk-cost evaluation is an appropriate approach to putting risk and economic factors on a unified scale of monetary value.

The risk cost can be evaluated using a unit energy interruption cost times the expected energy not supplied. However, quantification of the interruption cost is complex and immature. The customer damage functions that are based on customer surveys have been presented for years. Although this method is applied in risk evaluation, there have been a lot of debates about its use. One typical viewpoint is that the customer damage functions do not represent utility damages and it is inappropriate to use them to balance the capital investment of a utility. The opposite argument is that customers eventually pay all costs through rates and, thus, the customer's interruption cost should be considered in utility system planning. This is a complicated issue that relies on many nontechnical aspects, including regulation or deregulation, ownership, and rate design.

The following three methods of evaluating the unit interruption cost are discussed here. Different utilities may select a different method in their practice and studies.

1. Method based on customer damage functions. A customer damage function is obtained from customer surveys and the relevant statistical analysis. Based on a wide investigation of utility companies across Canada, for instance, the average unit interruption cost is between $4/KWh and $10/KWh. This is the average social damage cost due to electricity supply interruptions. It is important to recognize that the unit interruption cost is region-, country-, and system-specific. A utility should use the unit interruption cost that is based on its own customer survey and system analysis. This method has general implications and is particularly well suited to the utility with a customer-focused strategy.

2. Method based on capital investments. For a utility, any capital investment to reinforce the power system brings about an incremental decrease in system

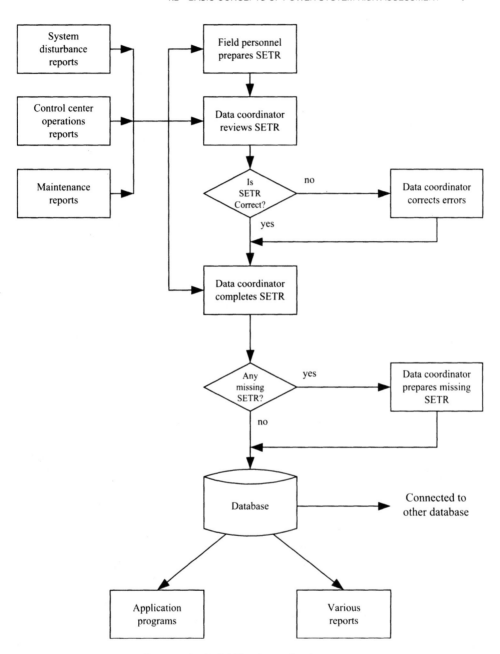

Figure 1.1 Reliability data collection system.

risk. In other words, there is a quantifiable relationship between the capital investment and the system risk index. With considerable studies in system reinforcement projects and relevant system risk assessments, the average unit interruption cost that is based on the capital investments can be obtained. The details of the method will be discussed in Chapter 12. This method may be particularly preferred by a private utility.

3. Method based on gross domestic product. A simple and useful method is to use the gross domestic product (GDP) concept. The gross domestic product for a province, state, or country divided by the total annual electric energy consumption of the province, state, or country results in a dollar value per kilowatt hour. This number reflects the average economic damage cost due to 1 KWh of electric energy loss in that province, state, or country. Obviously, this method is suitable for the utility owned by a government since the overall economic benefit in the province, state, or country must be considered as a whole.

1.3 OUTLINE OF THE BOOK

Considerable effort has been devoted to power system risk assessment in the past three decades. There are many technical papers and several books in this area [1, 3, 5–8, 19, 20]. The majority of the books have focused on analytical techniques [3, 5–7] and there has been one book on Monte Carlo simulation methods [1]. The emphasis in these books is placed on the theory and general methods for power system reliability evaluation, although some examples using small test systems have been used for demonstration purposes. This book has the following features:

- Most of this book is devoted to practical applications. The topics addressed are the actual issues in the utility industry. The special concepts, methods, and procedures for the different applications in planning, operation, maintenance, and asset management are developed. The examples provided are from real utility systems. Particularly, the results in the applications have been implemented in the utility's practice.

- The outage models of system components are systematically discussed in Chapter 2. Some concepts are new, such as the semiforced outage and individual model for common cause outages. Aging failure modeling is presented and applied to power system risk assessment for the first time.

- One whole chapter (Chapter 3) is committed to the parameter estimation of component outage models. This is an issue associated with how to perform statistical processing for historical outage records to handle the uncertainty of input data. It has not been sufficiently addressed so far in power system risk assessment.

- In evaluation methods, both the analytical and Monte Carlo approaches are discussed, with an emphasis on the applied techniques and actual considerations in generation, transmission, substation, and distribution systems. The methods that are not popular in practical applications are not included.

The book can be divided into three parts. The first part is the modeling and methods that are discussed in Chapters 2 to 5. The second part, Chapters 6 to 13, focuses on the eight topics in practical applications. The third part consists of the five appendices, which provide the background knowledge needed in power system risk assessment.

Chapters 2 and 3 discuss modeling and data issues at the system component level. Chapter 2 presents the outage models of system components, including various independent and dependent outages. These are the basis of system risk evaluation. Chapter 3 discusses the estimation methods of the parameters used in component outage models. These include point and interval estimations of failure data, experimental distributions of failure statistics, and parameter estimation in aging failure models. The essence of parameter estimation is the determination of input data in risk evaluation. This is one of the key steps in actual applications.

Chapters 4 and 5 discuss the evaluation methods and analysis techniques for overall systems. Chapter 4 summarizes the general risk evaluation methods that have been extensively applied in power system risk assessment. The probability convolution, series and parallel networks, Markov equations, and frequency-duration approaches for simple systems are described first, then emphasis is placed on the state enumeration and Monte Carlo simulation methods for complex systems. The merits and limitations of the different methods are discussed from the viewpoint of practical engineering application.

Chapter 5 illustrates the risk evaluation methods for generation, distribution, substation, and transmission systems using both the Monte Carlo simulation and state enumeration techniques. The focus is on the approaches of selecting system states and the techniques for performing system analyses. The practical considerations in the applications are discussed. The formulas of the risk indices for generation, distribution, substation, and transmission systems are derived. The formulas have different expressions for the Monte Carlo simulation and state enumeration techniques.

Eight application topics are presented in Chapters 6 to 13. In each chapter, not only are the concept, method, and procedure for a specific application described, but examples are also given to demonstrate the details of each application.

Chapter 6 illustrates the application of system risk assessment in transmission development planning, which is often called probabilistic planning in the utility environment. After the concept and approach are discussed, two examples are provided. The first one shows how to determine the lowest cost planning alternative and the second one shows how to apply different planning criteria.

Chapter 7 addresses the application of system risk assessment in transmission operation. The special aspects associated with operation modes are discussed and these include the impacts on system risk of load transfers, generation pattern changes, network reconfigurations, and switching actions. In addition to the example of determining the lowest risk operation mode in a large system, the simple case of using hand calculations is also offered. This provides the particulars of the evaluation procedure, which can be performed on a spreadsheet.

Chapter 8 discusses the application of system risk assessment in generation source planning. The basic concept and method for generation reliability planning under the network constraint condition are explained and two examples are demonstrated. One is associated with selection of the location and size of a cogeneration plant and the other with decision making on the retirement of a local generation plant in a utility.

Chapter 9 presents the application of system risk assessment in subtransmission systems and substation configurations. The general method for risk evaluation in the combined system of transmission network and substations is developed. Two examples are given to illustrate how to use the presented method in selecting the best substation configuration under the transmission network constraint and in determining the optimal transmission line arrangement connected to multiple substation configurations, respectively.

In Chapter 10, reliability-centered maintenance is discussed. The application of risk assessment in this area has a wide range of implications. Following the discussion of the basic tasks in reliability-centered maintenance, three examples are provided. The first is determination of the lowest risk maintenance scheduling. The second addresses the issue of workforce planning in maintenance. The third shows that a reliability-centered maintenance problem in the real life does not always have to be complex and can be solved through simple calculations in some cases.

In Chapter 11, probabilistic spare equipment planning is addressed. This area has been a challenge in the electric power industry for years. The spare issue is tightly related to equipment aging and thus the aging failure mode has to be considered. The spare analysis methods based on the risk criteria and probabilistic cost models are presented. Two practical applications from the utility projects are discussed in detail. In the first one, the number and timing of spare transformers are determined, and in the second one, the rational on-line redundancy of 500 kV reactors is analyzed.

Chapter 12 discusses the special application of risk assessment in reliability-based transmission service pricing. In the deregulation environment of the power industry, system reliability becomes part of transmission services and must be reflected in the price design. The basic concept, calculation method, and rate design based on quantified system risk evaluation are illustrated. A utility example is used to explain the application.

Chapter 13 presents probabilistic transient stability assessment. This is a relatively immature area in power system risk evaluation. The failure models and simulation methods associated with transient stability are addressed. A utility system in which two applications are discussed is used as an example. One is associated with the calculation of probabilistic transfer limit and the other with the determination of probabilistic generation rejection.

There are five appendices in total. Appendices A and B contain the mathematical knowledge for risk assessment. Appendix A provides basic probability concepts and Appendix B presents the elements of Monte Carlo simulation. Appendices C and D give the fundamentals of the power flow models and the optimization algorithms, respectively, that are used in power system risk assessment. Appendix E provides the three probability distribution tables that are often utilized in parameter estimation.

This book does not pretend to include all the known and available materials on the subject. It focuses instead on the aspects that have not been but should be sufficiently addressed in power system risk assessment. It will enable the reader who lacks experience in the power utility industry, such as a university student, to gain the practical ideas required in performing risk evaluation of real power systems. It will also help the reader, such as an engineer, who wants to learn more in modeling and methods and apply them to solve his or her actual problems, to acquire the requisite expertise in power system risk assessment.

CHAPTER 2

OUTAGE MODELS OF
SYSTEM COMPONENTS

2.1 INTRODUCTION

This chapter systematically discusses the outage models of system components. A power system consists of various components, such as generators, lines, cables, feeders, transformers, breakers, switches, reactors, and capacitors. Generally, component outages can be divided into two categories: independent and dependent. The models of independent outages are described first in Section 2.2, and those of dependent outages in Section 2.3.

The category of independent outages can be further broken down. There are different classification methods. They can be classified as forced, semiforced, and planned outages in terms of outage nature, or as full and partial failures in accordance with failure states. Forced outages are generally distinguished as repairable and nonrepairable failures. Considerable failures are repairable, whereas nonrepairable failures also often take place in real life. Equipment aging is a general phenomenon and has gradually become a major concern in many utilities. Therefore, the aging failure mode of components has to be incorporated in risk evaluation when the components approach their end of life. In the traditional risk evaluation of power systems until now, only repairable failures were considered and aging failures were ignored. Excluding the aging failures will most likely result in underestimation of power system risk. This book treats the issue of aging failure modeling as one of its main focuses.

There are different kinds of dependent outages. A simultaneous failure of two overhead lines on the same tower due to lightning is a typical example. This type of failure has been called a common-cause outage. Any failure of terminal equipment at a substation may cause multiple line outages. An adverse weather condition or a major storm disaster greatly increases the failure possibility of all the components

Risk Assessment of Power Systems. By Wenyuan Li
ISBN 0-471-63168-X © 2005 the Institute of Electrical and Electronics Engineers, Inc.

exposed to it. Protection coordination in power systems brings a chance of cascading outages, although it also plays an important role in system security and safety. All these dependent outages generally produce much more severe consequences than independent outages. Major power outages and blackouts have historically been associated with dependent failure events.

2.2 MODELS OF INDEPENDENT OUTAGES

In this book, independent outage modes of power system components are categorized as follows:

- Forced outage
    Repairable forced failure
    Nonrepairable forced failure
        Aging failure (end-of-life failure)
        Chance failure
- Planned outage
- Semiforced outage
- Partial failure mode
- Multiple failure mode

A forced outage happens randomly and is totally out of one's control. On the other hand, a planned outage is not caused by a failure but is scheduled by personnel, such as a maintenance or replacement activity. There also exist some outages that lie between these two categories. For example, a cable with a serious oil leakage will not be forced to outage immediately but has to be taken out of service within a limited time period. This type of outage is called semiforced or semiplanned.

The majority of forced outages in a power system are repairable, whereas some fatal failure events can also take place. Nonrepairable failures can be further classified into aging and chance failures. The aging failures should be a mandatory consideration in risk evaluation of power systems with aged equipment. A nonrepairable chance failure is like an accident. Its probability of occurrence is extremely small and may not be necessary to be included in an evaluation in most cases.

The partial failure mode is related to the case in which a piece of equipment can be operated at one or more derated states. This leads to a multiple state model. A multiple failure mode refers to a unified model of multiple component independent outages. Essentially, this should be treated as a system. However, the model of two components is directly used in many simple cases without the need for using a method for system risk evaluation.

2.2.1 Repairable Forced Failure

Repairable forced failures can be modeled using a steady up–down–up cycle process. Figures 2.1 and 2.2, respectively, show the cycle process and its state tran-

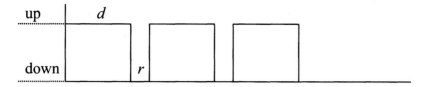

Figure 2.1 Up and down process of a repairable component.

sition diagram. Mathematically, the average unavailability in the long-term process is defined by one of the following three definitions [1-3]:

$$U = \frac{\lambda}{\lambda + \mu} = \frac{MTTR}{MTTF + MTTR} = \frac{f \times MTTR}{8760} \tag{2.1}$$

where λ is the failure rate (failures/year), μ the repair rate (repairs/year), $MTTR$ the mean time to repair (hours), $MTTF$ the mean time to failure (hours), and f the average failure frequency (failures/year). The three definitions are the same in essence. Only two of the parameters in Equation (2.1) are independent. In other words, only if two of them are known can the others be calculated. The following relationships can be obtained from Equation (2.1).

Let $d = MTTF/8760$ and $r = MTTR/8760$. Essentially, d and r are still $MTTF$ and $MTTR$, but in the unit of years. We have

$$\lambda = \frac{1}{d} \tag{2.2}$$

$$\mu = \frac{1}{r} \tag{2.3}$$

$$f = \frac{1}{d + r} \tag{2.4}$$

$$U = fr \tag{2.5}$$

$$f = \frac{\lambda}{1 + \lambda r} \tag{2.6}$$

$$\lambda = \frac{f}{1 - fr} \tag{2.7}$$

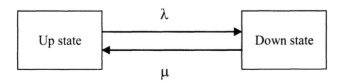

Figure 2.2 State space diagram of a repairable component.

It is important to stress that f and λ are two completely different quantities and this can be clearly seen from Equations (2.6) and (2.7). In most cases, both f (or λ) and r are small values and, therefore, f and λ are numerically close. However, r may have a large value in some special cases, such as the repair time for a submarine cable. In such a case, replacing λ by f or vice versa will create a relatively large error.

2.2.2 Aging Failure [25, 27]

An aging failure (end-of-life failure) is a nonrepairable failure. It may suddenly happen when a component enters the wear-out stage on the life basin curve, as shown in Figure 2.3. No repair time is associated with an aging failure since once a component fails due to aging, it fails forever. The aging failure is a conditional failure event that depends on the history; that is, how many years a system component has survived. Clearly, this condition cannot be modeled by Equation (2.1). Also, the aging failure is associated with the failure rate of increasing with time but not at the constant average failure rate that is used in Equation (2.1). In other words, the aging failure mode should be modeled in a different way.

2.2.2.1 Probability of Transition to Aging Failure. The probability of transition to aging failure of a component is a conditional probability. As shown in Figure 2.4, T is the age and t is a specified subsequent period to consider. The probability of transition to aging failure is defined as the probability that the component fails within t, given that it has survived until T. The survival probability is defined as the probability that the component still survives by the end of $T + t$, given that it has survived until T.

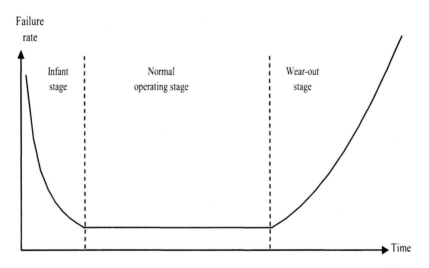

Figure 2.3 Basin curve of a component's life.

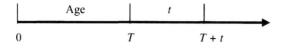

Figure 2.4   Aging failure concept.

The probability of transition to aging failure and the survival probability can be modeled using a posteriori Weibull or normal distribution [2]. Figure 2.5 shows the relationship between the failure rate and time for a normal distribution failure density function, and Figure 2.6 the same relationship for a Weibull distribution failure density function. It can be seen that the relationship shown in the two figures (for the Weibull distribution, see the curve with the shape parameter $\beta > 1$) is consistent with that expressed in the wear-out stage on the life basin curve in Figure 2.3.

According to the definition of reliability function and the conditional probability concept, the probability of transition to aging failure of a component in a subsequent period t after having survived for T years can be calculated by

$$P_f = \frac{\int_T^{T+t} f(t)dt}{\int_T^{\infty} f(t)dt} \tag{2.8}$$

where $f(t)$ is the failure density probability function (Weibull or normal distribution).

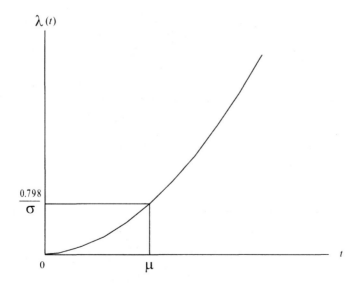

Figure 2.5   Relationship between $\lambda(t)$ and t for a normal distribution.

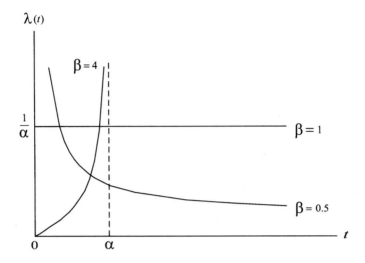

Figure 2.6   Relationship between $\lambda(t)$ and t for a Weibull distribution.

2.2.2.2   *Unavailability Due to Aging Failure.*

Although Equation (2.8) provides the probability of transition to aging failure, it is difficult to directly apply it in power system risk assessment. A power system may have been operating for many years and will be continuously operated. System components may have experienced many failures and repairs. What can be focused on is the average steady-state risk of the system. The risk evaluation methods for power systems that have been developed so far basically use the unavailability of components. The essence of Equation (2.1) is to calculate the average unavailability due to repairable failures over a long time. In order to incorporate the aging failure mode in a unified manner with the repairable failure mode, it is necessary to develop a model for the unavailability due to aging failures.

The unavailability of a component due to aging failure is defined as the probability that it is found to be unavailable during t, given that it has survived for T years. The component can fail at any point within t with a different probability. Therefore, the unavailability should be the conditional mathematical expectation of the time length not available during t divided by the period t. The condition is still the age T. The availability is 1.0 minus the unavailability.

It is obvious that the probability of transition to aging failure and the unavailability due to aging failure are two completely different concepts. The probability to aging failure is the probability of transition from a survival state to a failure state within t, whereas the unavailability is the average probability that a component is found not available during t. On the other hand, they have some common features. Both are defined by the conditional probability and associated with two time parameters: the age (T) and the subsequent time period to consider (t).

The unavailability of repairable failures given in Equation (2.1) is the unconditional average probability of being unavailable over the time and thus it is constant in the given time period. The unavailability of aging failures is associated with the age and a specified time period. However, the two unavailability concepts can be used in the same time frame, that is, the time period considered for the aging failure. This is due to the fact that we always have a specified period in power system risk evaluation. For instance, we often assess system risk on a yearly basis and calculate annual risk indices. We also have a reference year. Once the reference year (such as 2003) is selected, the age of each component whose aging failure is considered becomes known. The subsequent year is the time period to consider. In the system risk evaluation only considering repairable failures, although we do not explicitly emphasize the concept of the "subsequent" year, we do calculate annual indices based on the system conditions (system configuration, load level, and generation pattern) for that year.

In order to develop the unavailability of aging failures, the subsequent period to consider (i.e., t) is divided into N equal intervals with the interval length Δx. The endpoints of the intervals are marked by t_i ($i = 0, 1, 2, \ldots, N$), as shown in Fig. 2.7. By applying Equation (2.8), the probabilities to aging failure in the subsequent subperiods between t_0 and t_1, t_0 and t_2, $\ldots$, t_0 and t_N, given that it has survived until T (i.e., point $t_0 = 0$), can be calculated. Denoting these probabilities by P_{fi}, we have

$$P_{fi} = \frac{\int_T^{T+t_i} f(t)dt}{\int_T^\infty f(t)dt} \qquad (i = 1, 2, \ldots, N) \qquad (2.9)$$

The failure probabilities within each interval therefore can be obtained from Equation (2.9). Let P_i ($i = 1, 2, \ldots, N$) denote the failure probability in the ith interval. The failure probabilities in all the intervals can be calculated through the direct difference between two consecutive P_{fi} ($i = 1, 2, \ldots, N$):

$$P_i = \frac{\int_T^{T+i\Delta x} f(t)dt - \int_T^{T+(i-1)\Delta x} f(t)dt}{\int_T^\infty f(t)dt} \qquad (i = 1, 2, \ldots, N) \qquad (2.10)$$

It has been assumed that the failure probability at any point within the ith interval is the same (i.e., P_i). This assumption will not create a significant error as long as the length of intervals (i.e., Δx) is sufficiently short or, say, N is large enough. In fact,

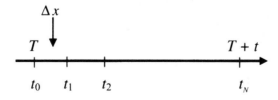

Figure 2.7 Intervals in the subsequent period t.

there is no error at all if we see this from a viewpoint of the limit when N approaches infinity. This is equivalent to changing Equation (2.12) below to the continuous integral expression.

Let us look at the first interval. If the component fails at the beginning of the interval (at point t_0), its available duration is zero. If it fails at the end of the interval (at point t_1), its available duration is Δx. Based on the assumption of equal failure probability at all points within the interval, the average available duration should be $\Delta x/2$. Therefore, the average unavailable duration within t is $(t - \Delta x/2)$ if it fails in the first interval. By applying a similar concept to any interval, we can derive the general equation of the average unavailable duration within t as follows:

$$UD_i = t - (2i - 1)\Delta x/2 \qquad (i = 1, 2, \ldots, N) \qquad (2.11)$$

where UD_i is the average unavailable duration within t when the component fails in the ith interval and Δx is the length of each interval.

With the failure probability and the average unavailable duration for each interval, the unavailability of a component in the specified subsequent period t is

$$U_a = \sum_{i=1}^{N} P_i \cdot UD_i/t \qquad (2.12)$$

The availability is

$$A_a = 1.0 - U_a \qquad (2.13)$$

We often use the annual unavailability due to aging failure in power system planning. In this case, t is one year. It should be kept in mind that the unavailability has to be recalculated for each year in the planning time frame since the age T, as its condition, increases every year. In principle, it is better to use the "functional age" in the model, which reflects the degree of "wear-out." The functional age depends on the usage history, maintenance situation, and actual status of equipment. Unfortunately, it is difficult to determine the functional age. In most cases, the natural age of equipment is used and can still produce an acceptable and reasonable result for system planning purposes.

2.2.2.3 Explicit Expressions of Equation 2.10. If Equation (2.10) has an explicit expression, the unavailability due to aging failure can be directly calculated from Equation (2.12).

If the aging failure is modeled by a posteriori normal distribution, the integration in Equation (2.10) does not have an accurate explicit expression. An approximation can be used as follows:

$$P_i = \frac{Q\left(\dfrac{T + (i-1)\Delta x - \mu}{\sigma}\right) - Q\left(\dfrac{T + i\Delta x - \mu}{\sigma}\right)}{Q\left(\dfrac{T - \mu}{\sigma}\right)} \qquad (i = 1, 2, \ldots, N) \qquad (2.14)$$

where μ and σ are the mean and standard deviation of the normal distribution and the function Q is approximated by

$$Q(y) = \begin{cases} w(y) & \text{if } y \geq 0 \\ 1 - w(-y) & \text{if } y < 0 \end{cases}$$

$$w(y) = z(y)(b_1 s + b_2 s^2 + b_3 s^3 + b_4 s^4 + b_5 s^5)$$

$$z(y) = \frac{1}{\sqrt{2\pi}} \exp\left(-\frac{y^2}{2}\right)$$

$$s = \frac{1}{1 + ry}$$

$$r = 0.2316419$$

$$b_1 = 0.31938153$$

$$b_2 = -0.356563782$$

$$b_3 = 1.781477937$$

$$b_4 = -1.821255978$$

$$b_5 = 1.330274429$$

If the aging failure is modeled by a posteriori Weibull distribution, Equation (2.10) can become

$$P_i = \frac{\exp\left[-\dfrac{T + (i-1)\Delta x}{\alpha}\right]^\beta - \exp\left[-\dfrac{T + i\Delta x}{\alpha}\right]^\beta}{\exp\left[-\dfrac{T}{\alpha}\right]^\beta} \qquad (i = 1, 2, \ldots, N) \quad (2.15)$$

where α and β are the scale and shape parameters for the Weibull distribution.

2.2.3 Nonrepairable Chance Failure

A nonrepairable chance failure refers to a random fatal failure in the normal operating stage of the life basin curve. Obviously, it corresponds to a constant failure rate and therefore can be modeled using an exponential distribution.

The conditional probability concept expressed in Equation (2.8) applies to not only the wear-out region but also the normal operating period. The failure density function for an exponential distribution is

$$f(t) = \lambda_h e^{-\lambda_h t} \tag{2.16}$$

where λ_h is the failure rate of a nonrepairable chance failure. Substituting Equation (2.16) into Equation (2.8), we obtain the probability of transition to the nonrepairable chance failure state:

$$P_f = \frac{\int_T^{T+t} f(t)dt}{\int_T^{\infty} f(t)dt} = \frac{\int_T^{T+t} \lambda_h e^{-\lambda_h t}dt}{\int_T^{\infty} \lambda_h e^{-\lambda_h t}dt} = \frac{e^{-\lambda_h T} - e^{-\lambda_h(T+t)}}{e^{-\lambda_h T}} = 1 - e^{-\lambda_h t} \tag{2.17}$$

It is interesting to note that T has disappeared from the equation. This is due to the no-memory feature of exponential distribution. In other words, the failure probability modeled by an exponential distribution does not depend on the age of components but only on the subsequent time length to consider. This is apparently correct since what we model is an accidental failure in the normal operating period.

The concept of the unavailability due to nonrepairable chance failure is the same as that due to aging failure and, therefore, all equations in Section 2.2.2.2 can apply. The explicit expression to calculate P_i in Equation (2.12) for a nonrepairable chance failure is yielded by substituting Equation (2.16) into (2.10):

$$P_i = e^{-\lambda_h(i-1)\Delta x} - e^{-\lambda_h i \Delta x} \tag{2.18}$$

It must be emphasized that λ_h here is different from λ in Equation (2.1). λ_h is for the nonrepairable chance failure mode, whereas λ is for the repairable failure mode. As mentioned earlier, a nonrepairable chance failure is normally not considered in most cases since its failure probability or unavailability is generally very small and negligible for power system components. However, the concept has been discussed and may be used in some special cases.

2.2.4 Planned Outage

A planned outage may be necessary for reasons such as maintenance, replacement, refurbishment, or an operational requirement. Two modeling approaches can be applied for a planned outage. The first one is to assume that the time to planned outage and the time to recovery follow given distributions. The parameters of the distributions are estimated from historical records of planned outage activities. The second one is to recognize planned outages as the scheduled events that occur at prespecified intervals.

In the first approach, a planned outage is treated as a random event. Figure 2.8 shows the state space diagram for a forced failure and a planned outage. Applying the Markov method (see Section 4.2.3) to the state space diagram, we can obtain the following results:

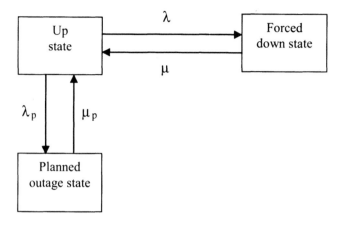

Figure 2.8   State space diagram for both forced and planned outages.

$$P_{up} = \frac{\mu_p \mu}{\lambda_p \mu + \lambda \mu_p + \mu_p \mu} \tag{2.19}$$

$$P_{fo} = \frac{\lambda \mu_p}{\lambda_p \mu + \lambda \mu_p + \mu_p \mu} \tag{2.20}$$

$$P_{po} = \frac{\lambda_p \mu}{\lambda_p \mu + \lambda \mu_p + \mu_p \mu} \tag{2.21}$$

$$f_p = \frac{\lambda_p \mu_p \mu}{\lambda_p \mu + \lambda \mu_p + \mu_p \mu} \tag{2.22}$$

$$f = \frac{\lambda \mu_p \mu}{\lambda_p \mu + \lambda \mu_p + \mu_p \mu} \tag{2.23}$$

Here, P_{up}, P_{fo}, and P_{po} are the probabilities of the up, forced outage, and planned outage states. f_p and f are the frequencies of the planned and forced outage states (outages/year). λ_p and λ are the transition rates of the planned and forced outage states. μ_p and μ are the recovery (repair) rates from the planned and forced outage states (repairs/year).

In most outage data collection systems, the outage rates (λ_p and λ) are not directly collected. Rather, the outage frequencies and repair times of components are recorded. Strictly speaking, therefore, we have to use Equations (2.22) and (2.23) to calculate λ_p and λ first before we can use Equations (2.19) to (2.21) to obtain the state probabilities. This causes complexity in input data preparations. From the viewpoint of computational approximation, it can be assumed that a planned outage of a component is not mutually exclusive with its forced outage. With this assumption, the forced and planned outages are represented using two separate two-state models—the forced outage by Figure 2.2 and the planned outage by Figure 2.9. Thus, the following model, similar to Equation (2.1), can apply to the planned outage:

$$U_p = \frac{\lambda_p}{\lambda_p + \mu_p} = \frac{MTTR_p}{MTTF_p + MTTR_p} = \frac{f_p \times MTTR_p}{8760} \qquad (2.24)$$

where λ_p, μ_p, and f_p are defined as earlier. $MTTF_p = 8760/\lambda_p$ is the mean time to planned outage (hours) and $MTTR_p = 8760/\mu_p$ the mean time to recovery (hours). U_p is the unavailability due to planned outage and has the same significance as P_{po} in Equation (2.21).

Obviously, the relationships similar to Equations (2.2) to (2.7) should hold as long as a subscript p representing a planned outage is assigned to each parameter. All the parameters are the average values based on historical records of planned outages.

The above assumption can greatly simplify data preparations and calculations in risk evaluation and it does not create an effective error in most cases. By using the combined model shown in Figure 2.8, the total outage probability is

$$U_t = P_{fo} + P_{po} = \frac{\lambda\mu_p + \lambda_p\mu}{\lambda_p\mu + \lambda\mu_p + \mu_p\mu} \qquad (2.25)$$

If two separate models for the forced and planned outages shown in Figures 2.2 and 2.9 are used, the total outage probability is:

$$U_t = 1 - \left(\frac{\mu_p}{\mu_p + \lambda_p}\right)\left(\frac{\mu}{\mu + \lambda}\right) = \frac{\lambda\mu_p + \lambda_p\mu + \lambda_p\lambda}{\lambda_p\mu + \lambda\mu_p + \mu_p\mu + \lambda_p\lambda} \qquad (2.26)$$

It can be seen by comparing Equation (2.25) with (2.26) that the error in Equation (2.26) is only associated with the term of $\lambda_p\lambda$ in both the denominator and numerator. This error is of the second order of magnitude and negligible since we know $\lambda_p \ll \mu_p$ and $\lambda \ll \mu$. It should be appreciated, however, that if there is a very long planned outage duration, the error due to the assumption will be increased. In this case, this approximation should be used with caution.

In the second approach, a planned outage is a prespecified schedule but not a random event. For example, say a transformer is scheduled to be out of service for maintenance from April 12 to May 4, 2003. In this case, the planed outage is defined by its

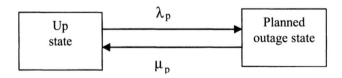

Figure 2.9 State space diagram for separate planned outage.

start and end dates and cannot be modeled using a transition state space diagram. It can be easily modeled using the Monte Carlo method, in which the system component is just simply taken out during the planned outage in the simulation process.

Both approaches have merits and demerits. The first approach cannot accurately reflect a planned schedule but is appropriate to use when any planned schedule is not known in long-term system planning. The second one can be applied in short-term operation planning where we either know start and end dates of a planned outage or want to find out the best start and end dates through a risk evaluation method.

2.2.5 Semiforced Outage

A semiforced outage refers to the case in which a physical problem of a system component will cause an outage with a time delay that depends on a forced reason but can be scheduled. The oil leakage of a cable or transformer is an example. The oil leakage generally does not create an immediate failure but requires an outage within a limited time. The time delay for the outage depends on several factors, including the severity of leakage, availability of manpower for repair, possibility of reducing the leakage by operational measures, and impact of the leakage on the system security and environment. In general, semiforced outages are not considered in the risk assessment for long-term system planning but are included in the risk evaluation for short-term operation planning.

Figure 2.10 shows the state space diagram for a semiforced outage. This is only an approximate representation since factors impacting the length of time delay cannot be precisely included in the model. Applying the Markov method (see Section 4.2.3) to the model, we have the following results:

$$P_{su} = \frac{\mu_s \mu_{so}}{\lambda_s \mu_s + \lambda_s \mu_{so} + \mu_s \mu_{so}} \tag{2.27}$$

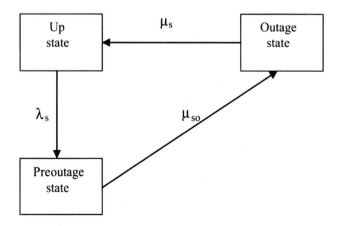

Figure 2.10   State space diagram for a semiforced outage.

$$P_{sp} = \frac{\lambda_s \mu_s}{\lambda_s \mu_s + \lambda_s \mu_{so} + \mu_s \mu_{so}} \tag{2.28}$$

$$P_{so} = \frac{\lambda_s \mu_{so}}{\lambda_s \mu_s + \lambda_s \mu_{so} + \mu_s \mu_{so}} \tag{2.29}$$

$$f_s = \frac{\lambda_s \mu_s \mu_{so}}{\lambda_s \mu_s + \lambda_s \mu_{so} + \mu_s \mu_{so}} \tag{2.30}$$

The subscript s indicates the quantity related to "semiforced." P_{su}, P_{sp}, and P_{so} are the probabilities of the up, preoutage (such as an oil leakage problem), and outage states. f_s is the frequency of the problem taking place (occurrences/year). λ_s is the transition rate from the up state to the preoutage state. μ_s is the repair rate, which is the reciprocal of the repair time, and μ_{so} is the transition rate from the preoutage state to the outage state, which is the reciprocal of the time delay. Note that μ_{so} is of the same order of magnitude as μ_s and much larger than λ_s because both the repair time and time delay are much shorter than the operating duration.

The average value for f_s, the repair time, and the time delay can be estimated based on historical records or from an engineering analysis. λ_s is calculated first using Equation (2.30) and then the three state probabilities can be obtained using Equations (2.27) to (2.29).

If the sequential Monte Carlo technique (see Section 4.3.3) is used, semiforced outages can be more accurately simulated because all factors impacting the length of time delay to outage can be included in simulations. The time delay can even be partly scheduled. For instance, if a semiforced outage does not cause any loss of load in the system, it can be scheduled as early as possible; otherwise, it should be delayed until other measures are ready, to avoid any load curtailment. However, it is important to appreciate that this schedule is different from a planned outage since there is a limited range for the time delay, which relies on the situation of the problem (such as oil leakage). Also, a semiforced outage cannot be treated as a deterministic event because the problem occurs randomly.

2.2.6 Partial Failure Mode

When a nonsevere failure occurs, some components (such as generating units and HVDC transmission lines) can still be operated in a derated state. This is called the partial failure mode. In order to include partial failures, a component can be represented by a multiple state model. Take a three-state model as an example. As shown in Figure 2.11, there is the derated state in addition to the full-up and full-down states.

In real life, it is difficult to obtain the data of transition rates between derated and full-down states as a repair is generally performed to make the equipment back to the full-up state but not to the derated state. It is a reasonable assumption to ignore the transitions between the derated and full-down states in most practical applications. With this assumption, the state space diagram becomes the one shown in Figure 2.12.

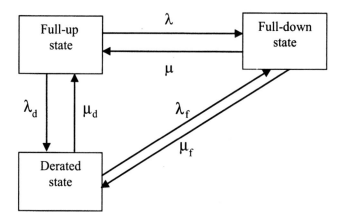

Figure 2.11 Three-state model with a derated state.

From a viewpoint of pure mathematical calculations, the simplified model is the same as the one shown in Figure 2.8, although they represent the completely different definitions. Therefore, Equations (2.19) to (2.23) can apply as long as the subscript p for the planned outage state in the equations is replaced by the subscript d to indicate the "derated state." Similar to what has been described for a planned outage in Section 2.2.4, the further simplification is to treat the derated failure and full forced outage as two separate and independent two-state models. This is equivalent to ignorance of the mutual exclusiveness between the derated and full-down states. It should be kept in mind that the reason why we can do the simplification is just because it does not create effective errors in the numerical results but not because the mutual exclusiveness does not exist. The simplification should not be used if there may be a relatively large error in some cases.

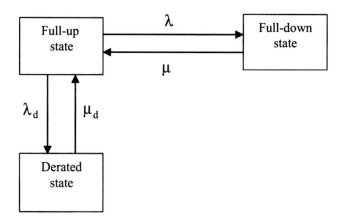

Figure 2.12 Simplified three-state model with a derated state.

2.2.7 Multiple Outage Mode

When multiple independent outages are considered, the union concept can be applied. Given two independent outages, the parameters of an equivalent outage are calculated using the following equations:

$$U_e = U_1 + U_2 - U_1 U_2 \tag{2.31}$$

$$\lambda_e = \lambda_1 + \lambda_2 \tag{2.32}$$

$$r_e = \frac{\lambda_1 r_1 + \lambda_2 r_2 + \lambda_1 r_1 \lambda_2 r_2}{\lambda_1 + \lambda_2} \tag{2.33}$$

$$f_e = f_1(1 - f_2 r_2) + f_2(1 - f_1 r_1) \tag{2.34}$$

where U, λ, r, and f represent the unavailability, failure rate, repair time, and failure frequency, respectively. The equations are the same as those for the series reliability network modeling and the derivation is given in Section 4.2.2.1. This set of equations can be repeatedly applied for a case with more than two outages. The first two outages are combined into one equivalent outage, and the resulting equivalent outage and the third one are combined into the second equivalent one, and so on.

If a nonrepairable aging or chance failure is one of the multiple outages, Equation (2.31) can still be applied but the other three equations cannot. This is because a nonrepairable failure does not have the concepts of failure frequency and repair time. A data collection system often classifies repairable forced failures in terms of failure causes. The above equations can also be used to calculate the total equivalent parameters from those for different failure causes. For instance, failure data of transmission components are generally categorized into two parts: component-related and terminal-related. When terminal equipment is not explicitly modeled, as an approximate modeling approach, terminal-related failure frequency and repair time of a primary component (a line, cable, or transformer) can be combined with its component-related failure frequency and repair time to obtain the total equivalent failure parameters using Equations (2.31) to (2.34).

It is worthy to note that if Monte Carlo methods are used to simulate multiple outages, there are two options. In the first one, Equations (2.31) to (2.34) are used to obtain the total equivalent failure parameters and then a random number is created to simulate the equivalent outage. In the second one, separate random numbers are utilized to simulate each of the outages and impacts of all the outages can be automatically captured in the simulation process.

2.3 MODELS OF DEPENDENT OUTAGES

Dependent outages in power systems are as popular as independent outages. They can be classified as follows:

- Common-cause outage
- Component-group outage
- Station-originated outage
- Cascading outage
- Environment-dependent failure

The common characteristic of all the dependent outage modes is that an outage state includes more than one component failure. The probability of a dependent outage event associated with multiple components is much larger than the probability of simultaneous independent outages of the same components. Therefore, dependent outages generally make greater contributions to system risk than independent outages.

2.3.1 Common-Cause Outage

A common-cause outage refers to simultaneous outages of multiple components due to a common cause. A typical example is the outage of two circuits on the same tower caused by a tower failure or lightning.

2.3.1.1 Composite Model. The traditional modeling approach for a common-cause outage is to combine it with independent outages into a composite state space model [1, 3]. Figure 2.13 shows the example of two components. Such a composite model has been widely used for years. However, it has the following disadvantages:

- The composite model implies the assumption that an independent failure and a common cause failure are mutually exclusive. If Figure 2.13 expresses the model for two circuits on the same tower, for instance, there is no state to represent the situation in which an independent failure of the circuits occurs at the same time as the tower failure. However, they can happen at the same time conceptually.
- When a common-cause outage is associated with more than two components, the composite model becomes very complex. The Markov equations for the model must be incorporated into the resolution process of system risk evaluation. The number of states in the model increases exponentially with the number of components. In other words, to model a case in which N components can be tripped out simultaneously by a protection action, $N^2 + 1$ state probability equations are needed only for this single common-cause outage and their independent failures. In the actual risk evaluation of most utility systems, multiple common cause outages have to be considered and the modeling of common-cause outages is only a small portion of the whole evaluation. This greatly increases the difficulty in programming and calculations. In fact, finding the analytical solution of the model for even two components shown in Figure 2.13 is not straightforward.

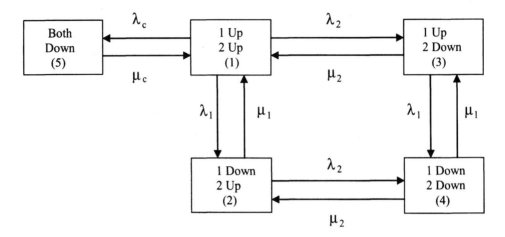

Figure 2.13   Composite model for independent and common-cause outages.

The first disadvantage is insignificant since the probability of concurrence of both common-cause and independent outages is normally negligible. It is more important to overcome the second disadvantage in actual risk assessments.

2.3.1.2 Individual Model [24].   A simple and accurate modeling approach is to use the individual two-state models for the common-cause outage and each independent outage, and the intersection concept for combinations of the outages. This idea is shown in Figure 2.14. The two-state model in Figure 2.14(a) is applied to independent outages of each component, and that in Figure 2.14(b) to their common-cause outage. Each independent outage and the common-cause outage are indepen-

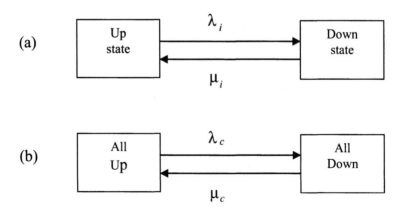

Figure 2.14   Individual models for independent and common-cause outages.

dent and not exclusive from each other. Thus, Equations (2.1) to (2.7) can be applied to each outage event, including the common-cause outage. The individual models can be summarized using Equations (2.35) to (2.40):

$$P_{iD} = \frac{\lambda_i}{\lambda_i + \mu_i} \qquad (i = 1, \ldots, n) \tag{2.35}$$

$$P_{iU} = \frac{\mu_i}{\lambda_i + \mu_i} \qquad (i = 1, \ldots, n) \tag{2.36}$$

$$\lambda_i = \frac{f_i}{1 - f_i r_i} \qquad (i = 1, \ldots, n) \tag{2.37}$$

$$P_{cD} = \frac{\lambda_c}{\lambda_c + \mu_c} \tag{2.38}$$

$$P_{cU} = \frac{\mu_c}{\lambda_c + \mu_c} \tag{2.39}$$

$$\lambda_c = \frac{f_c}{1 - f_c r_c} \tag{2.40}$$

The subscript i indicates the ith component and there are n components in total. The subscript c indicates the common cause outage of the n components. λ (failures/year) and μ (repairs/year) are the failure and repair rates. f and r are the failure frequency (failures/year) and repair time (years). r and μ are mutually reciprocal. P_{iD} and P_{iU} are the probabilities of the ith component in the down and up states. P_{cD} and P_{cU} are, respectively, the probabilities of the common-cause outage occurring and not occurring.

It can be seen that the equations for the independent and common-cause outages are the same in mathematical form. However, they have different meanings: when an independent outage takes place, only one component fails, whereas when the common-cause outage happens, all the n components fail. The individual modeling overcomes the two disadvantages of the composite model. First, any events, including the ones of concurrence of any independent outage and the common-cause outage, can be modeled. Second, any number of components in a common-cause outage can be easily and directly considered, using either an enumeration technique or Monte Carlo simulation approach (see Section 4.3).

It is important to appreciate that the "all up" state in the two-state model for the common-cause outage in Figure 2.14(b) has a completely different significance from the "1 up and 2 up" in the composite model in Figure 2.13. The probability of the "all up" state includes the portion of the common-cause outage not occurring but the independent component outage(s) occurring. The probability of the "1 up and 2 up" state in the composite model in Figure 2.13 is the probability of no outage (any independent or common-cause outage). In using the individual models, a combined state is defined as a combination of the common-cause and independent outages and its probability is calculated by

$$P_j = \prod_{i \subset su} P_{iU} \cdot \prod_{i \subset sd} P_{iD} \cdot \max\{kP_{cD}, (1-k)P_{cU}\} \qquad (2.41)$$

where P_j denotes the probability of any combined state j, su and sd are, respectively, the two sets of the components whose independent outage does not happen and happens in State j, and k is a variable having only two values of 0 or 1, with 0 indicating that the common cause outage does not happen in State j, and 1 otherwise. If su or sd is an empty set, the corresponding term is omitted.

2.3.1.3 Comparison between Composite and Individual Models.

The states associated with concurrence of the common-cause outage and any independent outage(s) are ignored in the composite model. Although such states can exist conceptually, their probability is very low and it can be qualitatively judged that there should not be a significant difference in the risk evaluation results obtained using the composite and individual models. A quantified proof for the case of two components is given as follows.

By applying the Markov method to the composite model, the probabilities of the five states shown in Figure 2.13 are obtained as follows:

$$P_1 = \frac{\mu_1 \mu_2 \mu_c}{(\mu_1 + \lambda_1)(\mu_2 + \lambda_2)\mu_c + \mu_1 \mu_2 \lambda_c} \qquad (2.42)$$

$$P_2 = \frac{\lambda_1 \mu_2 \mu_c}{(\mu_1 + \lambda_1)(\mu_2 + \lambda_2)\mu_c + \mu_1 \mu_2 \lambda_c} \qquad (2.43)$$

$$P_3 = \frac{\mu_1 \lambda_2 \mu_c}{(\mu_1 + \lambda_1)(\mu_2 + \lambda_2)\mu_c + \mu_1 \mu_2 \lambda_c} \qquad (2.44)$$

$$P_4 = \frac{\lambda_1 \lambda_2 \mu_c}{(\mu_1 + \lambda_1)(\mu_2 + \lambda_2)\mu_c + \mu_1 \mu_2 \lambda_c} \qquad (2.45)$$

$$P_5 = \frac{\mu_1 \mu_2 \lambda_c}{(\mu_1 + \lambda_1)(\mu_2 + \lambda_2)\mu_c + \mu_1 \mu_2 \lambda_c} \qquad (2.46)$$

For the individual models shown in Figure 2.14, Equations (2.35) and (2.36) are used for the independent outages of the two components, Equations (2.38) and (2.39) for their common-cause outage, and the intersection concept for the combinations of the outage events. The probability corresponding to State 1 in Figure 2.13 should be the probability that the independent outages of both the components and their common cause outage do not happen and is

$$P_1 = \left(\frac{\mu_1}{\mu_1 + \lambda_1}\right)\left(\frac{\mu_2}{\mu_2 + \lambda_2}\right)\left(\frac{\mu_c}{\mu_c + \lambda_c}\right)$$

$$= \frac{\mu_1 \mu_2 \mu_c}{(\mu_1 + \lambda_1)(\mu_2 + \lambda_2)\mu_c + \mu_1 \mu_2 \lambda_c + (\lambda_1 \lambda_2 + \mu_1 \lambda_2 + \mu_2 \lambda_1)\lambda_c} \tag{2.47}$$

Similarly, the probabilities corresponding to States 2 to 5 in Figure 2.13 are

$$P_2 = \left(\frac{\lambda_1}{\mu_1 + \lambda_1}\right)\left(\frac{\mu_2}{\mu_2 + \lambda_2}\right)\left(\frac{\mu_c}{\mu_c + \lambda_c}\right)$$

$$= \frac{\lambda_1 \mu_2 \mu_c}{(\mu_1 + \lambda_1)(\mu_2 + \lambda_2)\mu_c + \mu_1 \mu_2 \lambda_c + (\lambda_1 \lambda_2 + \mu_1 \lambda_2 + \mu_2 \lambda_1)\lambda_c} \tag{2.48}$$

$$P_3 = \left(\frac{\mu_1}{\mu_1 + \lambda_1}\right)\left(\frac{\lambda_2}{\mu_2 + \lambda_2}\right)\left(\frac{\mu_c}{\mu_c + \lambda_c}\right)$$

$$= \frac{\mu_1 \lambda_2 \mu_c}{(\mu_1 + \lambda_1)(\mu_2 + \lambda_2)\mu_c + \mu_1 \mu_2 \lambda_c + (\lambda_1 \lambda_2 + \mu_1 \lambda_2 + \mu_2 \lambda_1)\lambda_c} \tag{2.49}$$

$$P_4 = \left(\frac{\lambda_1}{\mu_1 + \lambda_1}\right)\left(\frac{\lambda_2}{\mu_2 + \lambda_2}\right)\left(\frac{\mu_c}{\mu_c + \lambda_c}\right)$$

$$= \frac{\lambda_1 \lambda_2 \mu_c}{(\mu_1 + \lambda_1)(\mu_2 + \lambda_2)\mu_c + \mu_1 \mu_2 \lambda_c + (\lambda_1 \lambda_2 + \mu_1 \lambda_2 + \mu_2 \lambda_1)\lambda_c} \tag{2.50}$$

$$P_5 = \left(\frac{\mu_1}{\mu_1 + \lambda_1}\right)\left(\frac{\mu_2}{\mu_2 + \lambda_2}\right)\left(\frac{\lambda_c}{\mu_c + \lambda_c}\right)$$

$$= \frac{\mu_1 \mu_2 \lambda_c}{(\mu_1 + \lambda_1)(\mu_2 + \lambda_2)\mu_c + \mu_1 \mu_2 \lambda_c + (\lambda_1 \lambda_2 + \mu_1 \lambda_2 + \mu_2 \lambda_1)\lambda_c} \tag{2.51}$$

Note that State 5 is the one in which the independent outages of both the components do not take place but only their common-cause outage occurs. The probability of any or both of the independent outages occurring at the same time as the common-cause outage, which has been ignored in the composite model, is associated with three states and can be calculated as

$$P_0 = \left(\frac{\lambda_1}{\mu_1 + \lambda_1}\right)\left(\frac{\mu_2}{\mu_2 + \lambda_2}\right)\left(\frac{\lambda_c}{\mu_c + \lambda_c}\right)$$

$$+ \left(\frac{\mu_1}{\mu_1 + \lambda_1}\right)\left(\frac{\lambda_2}{\mu_2 + \lambda_2}\right)\left(\frac{\lambda_c}{\mu_c + \lambda_c}\right)$$

$$+ \left(\frac{\lambda_1}{\mu_1 + \lambda_1}\right)\left(\frac{\lambda_2}{\mu_2 + \lambda_2}\right)\left(\frac{\lambda_c}{\mu_c + \lambda_c}\right) \tag{2.52}$$

$$= \frac{(\lambda_1 \lambda_2 + \mu_1 \lambda_2 + \mu_2 \lambda_1)\lambda_c}{(\mu_1 + \lambda_1)(\mu_2 + \lambda_2)\mu_c + \mu_1 \mu_2 \lambda_c + (\lambda_1 \lambda_2 + \mu_1 \lambda_2 + \mu_2 \lambda_1)\lambda_c}$$

It can be seen that the five state probabilities derived from the individual models have one more term in the denominator compared to those obtained from the composite model, which is $(\lambda_1\lambda_2 + \mu_1\lambda_2 + \mu_2\lambda_1)\lambda_c$. Since μ_1 or $\mu_2 \gg \lambda_1$, or λ_2 or λ_c, this term is extremely small compared to the rest of the denominator and thus negligible. In other words, the five state probabilities obtained from the two common-cause models are effectively equal in the numerical sense. For the same reason, the probability of concurrence of independent and common-cause outages P_0 is extremely low. Although the implied assumption in the composite model does not create an effective error in the result, the individual models avoid the use of Markov equations and are much simpler to calculate. It can be easily extended to consider any number of components in a common-cause outage mode. It also conceptually recognizes the fact that independent and common-cause outages are not mutually exclusive.

It is interesting to note the difference between the modeling ideas for the planned outage in Section 2.2.4 and for the common-cause outage in this section. Using the individual two-state model for a planned outage is approximate, as the mutual exclusiveness between planned and forced outages, which exits in real life, has been ignored. Using the individual two-state model for a common-cause outage is accurate since the concurrence (nonexclusiveness) of independent component outages and the common-cause outage is captured. On the contrary, the composite model using the combined state space diagram in Figure 2.13, in which the exclusiveness is assumed, is approximate. It is crucial in any outage modeling to have a full understanding of how failure events can happen.

2.3.2 Component-Group Outage

A component-group outage is defined as the case in which several components form a group and one component failure leads to simultaneous outages of all the components. Obviously, this is similar to the common-cause outage conceptually. The difference is that any component associated with the common-cause outage can also have its individual outage, whereas any component in the group outage does not, since all the components in the group always go to outage together. The model shown in Fig. 2.14(b) can be used for the group outage. The corresponding formulas are Equations (2.38) to (2.40). We do not need the model in Figure 2.14(a) because no single component outage event occurs in this case. It looks like the component-group outage is a specific example of the common-cause outage from a modeling viewpoint. However, it should be recognized that the physical mechanism is different. For a group outage, in most cases, only one of the components in the group really fails due to a fault, whereas the other components do not fail but go to outage because of a protective action.

There are two modeling approaches. The first one is to treat each component failure in a group separately. If there are N components in the group, we have N independent failure events and each one causes all the components to go to outage. When a probability distribution of the failure data for each component is considered, this approach is preferable. The second one is to calculate the equivalent failure frequency and repair time (or failure probability) for the whole group, using the

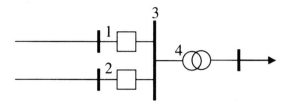

Figure 2.15 Example of a component-group outage.

series formula (see Section 2.2.7), and to use the model in Figure 2.14(b) only once. This second approach reduces efforts in calculations but can be used only when the average values of the failure data are employed. There may be a case in which only one component's failure in the group is considered and all other component failures are not taken into account in the group outage model. In some cases, a component can be modeled in such a way that it is a member of a group outage and, at the same time, its own failure is represented using another separate model.

Figure 2.15 presents a representative example of the group outage. This is a simple network with two lines to supply a load through a transformer. There is a breaker on the end of each line. Apparently, the two breakers (denoted by 1 and 2), transformer (denoted by 4), and busbar (denoted by 3) form a component group. A short-circuit on any of the four components will lead to their simultaneous outages, although only one is a real failure that is caused by the fault.

There may be T-connections on a line in transmission and distribution systems, as shown in Figure 2.16. These are often called the tapping points, where a subbranch or lateral feeder is directly connected to the line to draw power. In the power-flow model, the tapping points are treated as buses, as if they were the real ends (substations) of a line. In other words, a section between two T-connection points has been modeled as a branch in the same way as other lines without T-connections in power-flow calculations. In risk evaluation, however, a fault in one section will result in outages of all sections since they physically belong to one line without any isolator. Therefore, all the sections of one line in this case can be represented using the group-outage model while the power-flow model is still kept as usual. It should be noted that each section no longer has its own failure.

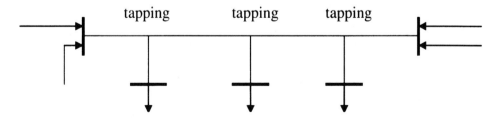

Figure 2.16 Line with T-connections modeled as a component-group outage.

2.3.3 Station-Originated Outage

A station-originated outage is a terminal-related outage of a primary component in transmission or distribution systems. It was mentioned in Section 2.2.7 that the terminal-related outage data for a primary component can be combined with the outage data of the component itself to obtain the equivalent total failure data. However, this is just an approximate approach. When a substation layout is considered, a more comprehensive model is needed. Figure 2.15 shows a simple case of a substation-originated outage that can be treated as a component-group outage. The practical situation is much more complicated and associated with the protection logic in substation configurations. Each case must be concretely analyzed.

Let us look at the example shown in Figure 2.17. The configuration is similar to that in Figure 2.15 except for the additional switch between the busbar and transformer. The added switch has caused a fundamental change for the impact of a fault in the transformer. Any grounding fault on the breaker 1 or 2, or the switch 5, or the busbar 3 will lead to the outage of all four components until the failed component is repaired. Thus, the four components can be represented using a group outage model. However, if a grounding fault happens in the transformer, breakers 1 and 2 are opened first, making themselves, together with the switch, busbar, and transformer, enter a temporary outage state. Then the switch is opened so that breakers 1 and 2 can be reclosed, allowing the two lines to be reconnected through them and the busbar. The switch and transformer will stay in the outage state until the fault is cleared or any damage caused by the fault is repaired so that the switch can be reclosed. The opening of the switch and reclosing of the breakers often require intervention of a substation operator, which will take some time. Obviously, the fault on the transformer cannot be modeled using a group outage but, rather, it can be represented using the three-state model shown in Figure 2.18. λ_{ts} denotes the failure rate due to a short-circuit fault, λ_{to} the failure rate due to an open circuit fault, μ_{tr} the repair rate, and μ_{sw} the switching rate. The significance of each state should be understood. The up or down status of each component in each state must be clearly identified. In the switching state, the two breakers, switch, busbar and transformer are in outage simultaneously. In the repairing state, only the switch and transformer are down. A set of equations formulating the relationship between state probabilities, frequencies, and transition rates can be obtained using the Markov method (see Section 4.2.3).

Different failure events are represented using varied models in a system risk evaluation process. For the configuration shown in Figure 2.17, for instance, the following failure events should be considered:

- The independent failures of the two lines
- The independent opened failures of the switch and two breakers
- The group-outage event representing a short-circuit fault on any of the two breakers, switch, and busbar
- The independent opened failure and short-circuit fault of the transformer modeled by Figure 2.18

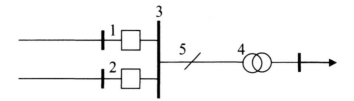

Figure 2.17 Example of a station-originated outage.

2.3.4 Cascading Outage

A cascading outage is defined as the event in which a failure of the first component causes a failure of the second one, and the second one's failure causes a failure of the third one, and so on. The first component is called the outage trigger. Such a cascading outage is often the main reason for a whole system blackout. A typical example is a 500 kV line in the direction parallel with a 230 kV line. When the 500 kV line fails, overloading occurs on the 230 kV line. If an appropriate operation measure is not taken in time, the overloading protection will trigger the 230 kV line outage, which may in turn create a wider overloading problem on other lines plus undervoltage at multiple buses. This may cause more system component outages, even resulting in a blackout.

Figure 2.19 shows the state space diagram for a cascading outage. It is a looped state transition process. State 0 corresponds to the normal state, State 1 to the failure of the trigger component, State 2 to the second component's failure, and so on. It has been assumed that all the failed components will return to the normal state at the same time. This is a general situation since only the trigger component may or may not need to be repaired and others could be brought back to service just by

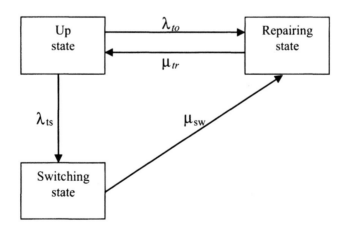

Figure 2.18 Three-state model for the transformer fault.

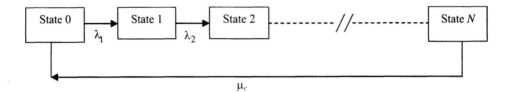

Figure 2.19 State space diagram for a cascading outage.

closing breakers in most cases. A set of equations for the model can be easily obtained using the Markov method. In an actual application, however, a reasonable approximation is ignorance of all the other states except States 0 and N. This is because the transition from a state to another in the cascading failure always happens so fast that no action can be carried out. Mathematically, we have $\lambda_1 \ll \mu_c \ll \lambda_i$ ($i = 2, \ldots, N$), where λ_1 is the failure rate of the trigger component, μ_c the recovery rate of the cascading outage that is usually the repair rate of the failed trigger component, and λ_i the ith transition rate in the cascading outage process. In other words, all the impacted components fail in a short time period as if they would fail simultaneously. This assumption allows use of the simple model between States 0 and N, which is similar to that for the common-cause outage shown in Fig. 2.14(b). All the components are up in State 0,whereas all of them are down in State N. Failure and repair rates between the two states are those of the trigger component. A basic difference is that the trigger component in the cascading outage no longer has its own independent outage, whereas all the components in the common-cause outage have their independent outages and there is no concept of trigger component. Essentially, this simplification enables us to model the cascading outage as a group outage.

2.3.5 Environment-Dependent Failure [1, 3]

Many system components are exposed to environments that can lead to an adverse or even catastrophic condition. Although this is generally infrequent and of short duration, the failure probability of components increases dramatically and overlapping failures of multiple components or even a whole subsystem failure can occur during this period. It should be noted that the failure process of overlapping outages in an adverse or catastrophic environment is conceptually different from a common-cause outage. In essence, the environment-dependent outage implies the dependence only between system components and the environment but not between components. Overlapping failures of components are still independent but are enhanced due to a common environment.

An adverse environment is usually an unfavorable weather condition like wind, rain, snow, etc., whereas a catastrophic environment could be any natural disaster such as a major ice storm, tornado, fire, flooding, or earthquake. It is difficult to develop an accurate model for the catastrophic environment since its probability of occurrence and the impact range can only be based on a rough estimation. However, once the estimation is accepted in the analysis, it can be reasonably assumed that all the components in the estimated range would simultaneously fail as if they were in a common-cause outage. The assumption enables us to perform a system risk assessment including the catastrophic environment impact. Although simplified and maybe inaccurate, this approach provides some quantified information that is useful in decision making.

For general weather conditions, the modeling approach can be better designed since historical records of weather are always available. Traditionally, weather is divided into two basic states: normal and adverse. The probabilities of the normal and adverse weather states can be calculated from the weather data. If the failure frequencies and repair times of components in both the normal and adverse states can be recognized, system risk is assessed separately for the two weather conditions and the resulting risk indices are weighted using the probabilities of the two weather states. However, most data collection systems do not distinguish between the failure events in the normal and adverse weather conditions but instead only respond to an average failure frequency and an average repair time in past years. In this case, Equations (2.55)–(2.58) below are used to calculate the failure frequencies and repair times for the two weather states.

According to the concept of mathematical expectation, we have

$$f_{to} = f_{ad} P_{ad} + f_{no}(1 - P_{ad}) \tag{2.53}$$

$$r_{to} = r_{ad} P_{ad} + r_{no}(1 - P_{ad}) \tag{2.54}$$

Here, f_{ad} and f_{no} are the failure frequencies in the adverse and normal weather conditions, f_{to} the average failure frequency, r_{ad} and r_{no} the repair times in the two weather conditions, and r_{to} the average repair time in the whole period. The P_{ad} and $(1 - P_{ad})$ are the probabilities of the adverse and normal weather conditions.

Two parameters of F and M can be introduced. F is the proportion of failures occurring in the adverse weather and it should be a value between 0 and 1. M is the multiple of the repair time in the adverse weather with respect to that in the normal weather and it must be equal to or larger than 1. Generally, F and M can be estimated from an engineering judgment. Otherwise, if the value of F or M is unknown, a sensitivity analysis can be performed. With the definitions of F and M, the following equations can be derived from Equations (2.53) and (2.54):

$$f_{ad} = \frac{f_{to} \cdot F}{P_{ad}} \tag{2.55}$$

$$f_{no} = \frac{f_{to}(1 - F)}{1 - P_{ad}} \tag{2.56}$$

$$r_{ad} = \frac{r_{to} \cdot M}{1 + (M - 1)P_{ad}} \tag{2.57}$$

$$r_{no} = \frac{r_{to}}{1 + (M - 1)P_{ad}} \tag{2.58}$$

Once the failure frequency and repair time are obtained, Equations (2.1), (2.3), and (2.7) are used to calculate the unavailability, repair rate and failure rate, for both the normal and adverse weather conditions.

In the case of overhead lines, it is possible that a line traverses two regions, where Region 1 is in the adverse weather and Region 2 in the normal weather. The series concept (see Section 4.2.2.1) can be applied to obtain the following equations:

$$\lambda_{le} = \lambda_{ad}R + \lambda_{no}(1 - R) \tag{2.59}$$

$$r_{le} = \frac{r_{ad}\lambda_{ad}R + r_{no}\lambda_{no}(1 - R) + r_{ad}r_{no} \cdot \lambda_{ad}R \cdot \lambda_{no}(1 - R)}{\lambda_{ad}R + \lambda_{no}(1 - R)} \tag{2.60}$$

$$U_{le} = U_{ad} + U_{no} - U_{ad}U_{no} \tag{2.61}$$

where

$$U_{ad} = \frac{\lambda_{ad}R}{\lambda_{ad}R + \mu_{ad}} \tag{2.62}$$

$$U_{no} = \frac{\lambda_{no}(1 - R)}{\lambda_{no}(1 - R) + \mu_{no}} \tag{2.63}$$

Here, λ (failure rate), r (repair time), U (unavailability), and μ (repair rate) have the same significance as defined in Section 2.2.1 except that the subscripts "*ad*" and "*no*" represent the adverse and normal weather conditions, respectively. R is the fraction of the line length in Region 1 that is exposed to the adverse weather environment.

Obviously, a similar concept can be extended to the cases in which an overhead line traverses more than two regions where different weather conditions are assumed.

2.4 CONCLUSIONS

The outage models of system components are the basis of power system risk assessment. This chapter has systematically discussed the details of various outage models.

In the independent outages, the two-state, repairable forced-outage model is of the most importance and has been extensively used in the traditional risk evaluation. The aging failure and semiforced outage are presented as the new modeling concepts. Equipment aging has become a critical factor, causing failures in many utility systems. In this case, ignoring aging failures will definitely lead to an underestimation of system risk. The unavailability due to aging failures and the unavailability due to repairable failures are conceptually compatible so that they can be used in the same manner. The semiforced outage concept is based on the actual need, such as the situation of a cable or transformer with oil leakage. In the engineering application, it can be represented using a three-state model. There are two modeling approaches for a planned outage. The one is to treat it as a random event, which can be represented using a three-state model or a simplified two-state model. The other is to consider it as a prescheduled event. The first approach is used in long-term development planning, whereas the second one is used in short-term operation planning. The partial failure mode not only applies to generating units but also to some transmission components like HVDC. In the actual application, the transitions between down and derated states are often ignored due to lack of data. This approximation generally does not create an effective error. The multiple-failure mode is essentially the representation of two or more independent failure components, which is obtained using the series network method. It is often used to derive the equivalent outage model from component-related and terminal-related failures.

In the dependent outages, the common-cause outage is the most popular one. There are two modeling methods for a common-cause outage: composite and individual models. The traditional composite model is theoretically approximate and requires relatively complicated calculations. The individual model is not only much simpler but also more accurate compared to the composite model. The component-group outage is addressed in this chapter. Conceptually, it is similar to but different from the common-cause outage. This concept can be used in modeling some substation configurations and tapping connections of transmission and distribution lines. The station-originated outage presents a general idea in modeling substation configurations. There is no generic model for all substation layouts. The key is to determine the failure events to be considered and analyze the concrete arrangements and protection logic in each event. The cascading outage is an important phenomenon in power system operation. This type of outage can be modeled using a reasonable approximation. The environment-dependent failure refers to the dependence between weather conditions and system component failures. It is crucial to collect historical weather statistics and develop the probabilistic relationship between the weather conditions and the failure parameters of system components.

An outage model can be either accurately developed or simplified in some cases. In designing or simplifying an outage model, it is important to understand how a failure event happens, what impact it has, and how a failed component recovers. It is extremely useful to have knowledge of the order of magnitude of each parameter in the models. This enables the analyst to make a rational decision in simplification.

CHAPTER 3

PARAMETER ESTIMATION IN OUTAGE MODELS

3.1 INTRODUCTION

Each outage model is characterized by parameters such as the failure frequency and repair time, failure rate and repair rate, transition rates between multiple states, unavailability, or mean life and standard deviation. The parameters are the input data in system risk evaluation and can be estimated from historical failure statistics.

For the repairable two-state model of a single component, only two parameters are needed and others can be calculated from them. Most data collection systems provide the failure frequency (or unavailability) and repair time. Generally, the failure rate and repair rate are not directly collected but can be calculated from the failure frequency and repair time. For the aging failure model, there is no concept of failure frequency and repair time. The definition of the unavailability due to aging failures was developed in Section 2.2.2. It has consistent significance with unavailability due to repairable failures so that both can be used in the same way in system risk evaluation. The primary parameters for the aging failure model are the mean life and standard deviation in the normal distribution model or the shape and scale parameters in the Weibull distribution model.

When an analytical method is used, a parameter in the outage model is the mean value based on historical statistics. In order to deal with the uncertainty in statistics, it is necessary to have a confidence range of the estimated parameter so that sensitivity studies can be performed. If Monte Carlo simulation is used, a probability distribution of the parameter can also be considered. In this case, we have to estimate both the mean and standard deviation for the assumed continuous distribution of the parameter or its experimental discrete distribution.

Risk Assessment of Power Systems. By Wenyuan Li
ISBN 0-471-63168-X © 2005 the Institute of Electrical and Electronics Engineers, Inc.

This chapter discusses the methods of parameter estimation, which are sometimes called the statistical models of parameters. In Sections 3.2 and 3.3, the point and interval estimations of mean and variance are described first as a general data processing approach. Then the four specific topics are presented. Section 3.4 addresses the estimation of the failure frequency of individual components and Section 3.5 the estimation of the probability from a binomial distribution. The experimental distribution concept of a parameter is illustrated in Section 3.6. The estimation of the parameters for the aging failure models is introduced as a newly developed method in Section 3.7.

3.2 POINT ESTIMATION OF MEAN AND VARIANCE OF FAILURE DATA

As mentioned, a parameter in the outage model is an average value of failure data. There are two average concepts: the average of sample data and the average over a given time period. For example, MTTR (mean time to repair) is the average of historical repair times of a single or multiple components. When a failure frequency is calculated based on multiple components, it can be the average of the failure frequencies of all individual components. However, the failure frequency of a single component is the average of failures over a given period. The sample mean is a popular estimation for the average of sample data and is discussed in this section. The estimation of the failure frequency of a single component will be addressed in Section 3.4.

3.2.1 Sample Mean

The arithmetic average of sample data expressed in Equation (3.1) is an unbiased estimate of the mean value:

$$\overline{X} = \frac{1}{n} \sum_{i=1}^{n} X_i \tag{3.1}$$

where, $\overline{X}$ is the mean, X_i a sample of a parameter, and n the number of sample data.

Although the arithmetic mean estimation is the most popular method, the geometric mean is also used in some cases. For instance, it is sometimes used to calculate the total mean from the mean values of subgroups of sample data. The geometric mean is defined as

$$\overline{X} = \left[\prod_{i=1}^{n} X_i \right]^{1/n} \tag{3.2}$$

In order to simplify the calculation, Equation (3.2) can be equivalently transformed into

$$\overline{X} = \ln^{-1}\left[\frac{1}{n}\sum_{i=1}^{n}\ln X_i\right] \qquad (3.3)$$

The weighting mean given in Equation (3.4) below is another important concept. In the case of multiple data sources or subgroups, such as the data from different geographical regions or different age groups, the weighting mean provides a more accurate estimate of the total mean:

$$\overline{X} = \sum_{i=1}^{M}\overline{X}_i W_i \qquad (3.4)$$

where $\overline{X}_i$ is the mean of the ith data subgroup, M is the number of data subgroups, and W_i is the dispersion dependent weighting factor for the ith data sub-group calculated by

$$W_i = \frac{1/\sigma_i}{\sum_{i=1}^{M}(1/\sigma_i)} \qquad (3.5)$$

where σ_i is the standard deviation for the ith data subgroup. The standard deviation of a discrete random variable is calculated as the square root of its sample variance (see Section 3.2.2.).

Table 3.1 shows the sample data of repair times for a generator and the comparison between the arithmetic and geometric means.

Table 3.2 shows the sample data of 230 kV line failure frequencies from three different sources. A comparison between the direct arithmetic mean and the weighting mean is given as follows.

Table 3.1 Sample data and sample mean of repair times for a generator

Repair	Time (h)
1	25.6
2	32.7
3	45.0
4	28.9
5	42.8
6	35.9
7	23.7
8	16.5
9	21.3
10	18.0
Arithmetic mean	29.04
Geometric mean	27.54

Table 3.2   Sample data of failure frequencies for 230 kV lines (failures/year/100 km)

Group 1	Group 2	Group 3
0.434	0.423	0.012
0.516	0.164	0.638
0.664	0.287	0.720
0.442	0.832	0.101
0.614	0.915	0.215
	0.092	1.041
		0.046

Using the direct arithmetic mean method, the total mean = 0.453. Using the weighting mean method:

Group 1: mean = 0.534; standard deviation = 0.092
Group 2: mean = 0.452; standard deviation = 0.316
Group 3: mean = 0.396; standard deviation = 0.372
The weighting factors: $W_1 = 0.65$, $W_2 = 0.19$, $W_3 = 0.16$
The total weighted mean = $0.534 \times 0.65 + 0.452 \times 0.19 + 0.396 \times 0.16 = 0.496$

3.2.2   Sample Variance

The variance is an indicator of the dispersion degree of a random variable. The standard deviation is the square root of the variance. In data processing, the sample variance s^2 of failure data is often used to replace the population variance σ^2. The sample variance is calculated by Equation (3.6):

$$s^2 = \frac{1}{n-1} \sum_{i=1}^{n} (X_i - \overline{X})^2 = \frac{1}{n-1} \sum_{i=1}^{n} X_i^2 - \overline{X}^2 \qquad (3.6)$$

where X_i, $\overline{X}$, and n are the same as defined in Section 3.2.1. It should be pointed out that although the denominator in Equation (3.6) is $n-1$ according to the strict definition of sample variance, it is often common to use n to replace $n-1$ in actual engineering calculations. Particularly, when n is large, using n or $n-1$ does not make a significant difference.

Table 3.3 presents an example of calculating the sample variance and sample standard deviation for the repair time data given in Table 3.1.

3.3   INTERVAL ESTIMATION OF MEAN AND VARIANCE OF FAILURE DATA

The methods described in Section 3.2 are used to estimate a single value of mean or variance and are called point estimation methods. It cannot be expected that a point

Table 3.3 Sample variance and sample standard deviation of repair times for a generator

Repair	Time (h)	$X_i - \bar{X}$	$(X_i - \bar{X})^2$
1	25.6	−3.44	11.8336
2	32.7	+3.66	13.3959
3	45.0	+15.96	254.7216
4	28.9	−0.14	0.0196
5	42.8	+13.76	189.3376
6	35.9	+6.86	47.0596
7	23.7	−5.34	28.5156
8	16.5	−12.54	157.2516
9	21.3	−7.74	59.9076
10	18.0	−11.04	121.8816
			883.9243

Sample mean = 29.04
Sample variance, 883.9/10 = 88.39
Sample standard deviation = 9.40

estimate hits a parameter exactly on the nose. It is desirable to have a confidence interval to cover the uncertainty of statistics. Such an interval contains the parameter to be estimated with the probability that is called the degree of confidence. The confidence interval is extremely useful for sensitivity studies in system risk evaluation.

3.3.1 General Concept of Confidence Interval [13, 16]

The sample mean $\bar{X}$ is not exactly the same as the real mean μ of whole population but just close to it. It can be easily imaged that a random interval exists to contain the real mean μ. Take the normal population $N(X; \mu, 1)$ (the mean μ unknown and the variance of 1) as an example. From the standard normal distribution table, we know that

$$\frac{1}{\sqrt{2\pi}} \int_{-1.96}^{1.96} e^{-(y^2/2)} dy = 0.95 \tag{3.7}$$

This implies that

$$p(-1.96 \leq (\bar{X} - \mu)\sqrt{n} \leq 1.96) = 0.95 \tag{3.8}$$

or

$$p\left(\bar{X} - \frac{1.96}{\sqrt{n}} \leq \mu \leq \bar{X} + \frac{1.96}{\sqrt{n}}\right) = 0.95 \tag{3.9}$$

where $\bar{X}$ is the sample mean of the normal population $N(X; \mu, 1)$ and the n is the number of sample data.

The random interval $[\overline{X} - (1.96/\sqrt{n}), \overline{X} + (1.96/\sqrt{n})]$ contains the unknown mean μ with probability of 0.95.

A general statement for the interval estimation is as follows. Assume that the distribution of a population variable X has an unknown parameter θ and we can build two functions of the variable's random sample sequence $\{X_i, i = 1, \ldots, n\}$: θ_1^* and θ_2^*, with $\theta_1^* \leq \theta_2^*$. If the following equation holds for a given α $(0 \leq \alpha \leq 1.0)$,

$$p(\theta_1^* \leq \theta \leq \theta_2^*) = 1 - \alpha \qquad (3.10)$$

then the interval (θ_1^*, θ_2^*) is a confidence interval of the parameter θ. The quantity $1 - \alpha$ is called the confidence degree and α is called the significance level, which is generally specified to be equal to or smaller than 0.05 (i.e., 5%).

It is important to comprehend the explanation for the confidence interval. The confidence degree is the probability that a random confidence interval contains the unknown parameter rather than the probability that the random parameter falls in a fixed interval. In other words, the confidence interval is varied depending on the sample size and how to build the two bound functions. However, for risk sensitivity studies in which a failure frequency or a repair time is considered to vary in a range, the confidence interval provides a good reference.

3.3.2 Confidence Interval of Mean

The following two theorems in statistics theory [16] enable us to estimate a confidence interval of the mean.

Theorem 1: If $\overline{X}$ and s are the sample mean and sample standard deviation of a random sample of size n from the normal population $N(X; \mu, \sigma^2)$, then $(\overline{X} - \mu)\sqrt{n}/s$ follows a t-distribution with $n - 1$ degrees of freedom.

According to Theorem 1, for the given significance level α, it can be affirmed that the random variable $(\overline{X} - \mu)\sqrt{n}/s$ is located between $-t_{\alpha/2}(n - 1)$ and $t_{\alpha/2}(n - 1)$ with probability of $1 - \alpha$, where $t_{\alpha/2}(n - 1)$ is such a value that the integral of the t-distribution density function with $n - 1$ degrees of freedom from $t_{\alpha/2}(n - 1)$ to ∞ equals $\alpha/2$. This concept is shown in Figure 3.1. Therefore, we have

$$-t_{\alpha/2}(n - 1) \leq \frac{\overline{X} - \mu}{s/\sqrt{n}} \leq t_{\alpha/2}(n - 1) \qquad (3.11)$$

or

$$\overline{X} - t_{\alpha/2}(n - 1)\frac{s}{\sqrt{n}} \leq \mu \leq \overline{X} + t_{\alpha/2}(n - 1)\frac{s}{\sqrt{n}} \qquad (3.12)$$

Theoretically, this confidence interval of μ can be used only for the normal population. However, it is a general practice to assume that failure data follow the normal

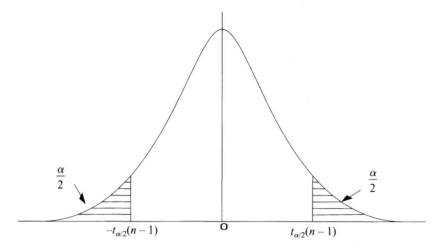

Figure 3.1 Confidence interval of mean using t-distribution.

distribution. An approximate confidence interval estimate of μ for a relatively large sample size from nonnormal distributions can be also obtained using the following Theorem 2.

Theorem 2: If $\overline{X}$ is the sample mean of a random sample of size n from the normal population $N(X; \mu, \sigma^2)$, then the sampling distribution of $\overline{X}$ is the normal distribution $N(X; \mu, \sigma^2/n)$.

When a random variable X does not follow a normal distribution, its sample mean $\overline{X}$ generally does not follow a normal distribution. For a large sample, however, any random variable X approximately follows a normal distribution according to the Central Limit Theorem. As a result, from Theorem 2, the sample mean $\overline{X}$ approximately follows the normal distribution $N(X; \mu, \sigma^2/n)$. Also, the sample variance s^2 is a consistence estimate of the population variance σ^2. By substituting the sample standard deviation s for σ, it can be affirmed that the random variable $(\overline{X} - \mu)\sqrt{n}/s$ is located between $-z_{\alpha/2}$ and $z_{\alpha/2}$ with the probability of $1 - \alpha$, where $z_{\alpha/2}$ is such a value that the integral of the standard normal density function from $z_{\alpha/2}$ to ∞ equals $\alpha/2$. This concept is shown in Figure 3.2. Therefore, we have

$$-z_{\alpha/2} \leq \frac{\overline{X} - \mu}{s/\sqrt{n}} \leq z_{\alpha/2} \tag{3.13}$$

or equivalently,

$$\overline{X} - z_{\alpha/2}\frac{s}{\sqrt{n}} \leq \mu \leq \overline{X} + z_{\alpha/2}\frac{s}{\sqrt{n}} \tag{3.14}$$

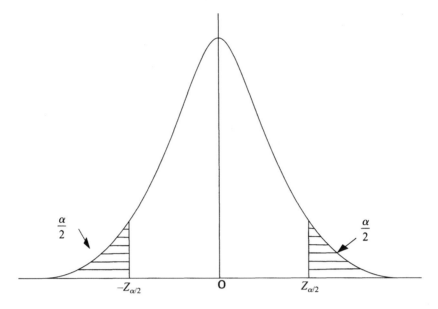

Figure 3.2   Confidence interval of mean using normal distribution.

It can be seen that the confidence interval estimate for a mean is not unique. When Equation (3.12) or (3.14) is used, $t_{\alpha/2}(n-1)$ or $z_{\alpha/2}$ can be obtained from the t-distribution or standard normal distribution table (see Appendix E). Generally, Equation (3.12) is used for a relatively small number of samples, whereas Equation (3.14) is used for a relatively large number of samples.

The following is a numerical example to show the interval estimation for the mean of the repair times given in Table 3.1 with a significance level $\alpha = 0.05$.

(1) Using the t-distribution. With $n = 10$ and $\alpha/2 = 0.025$, we can look up the value of $t_{\alpha/2}(n-1)$ from Table 2 in Appendix E, which is 2.262. The sample mean and sample standard deviation are $\overline{X} = 29.04$ and $s = 9.4$. The confidence interval is:

Lower limit: $29.04 - 2.262 \times (9.4/\sqrt{10}) = 22.32$
Upper limit: $29.04 + 2.262 \times (9.4/\sqrt{10}) = 35.76$

(2) Using the normal distribution with s to replace σ. With $n = 10$ and $\alpha/2 = 0.025$, we can look up the value of $z_{\alpha/2}$ from Table 1 in Appendix E, which is 1.96. The confidence interval is:

Lower limit: $29.04 - 1.96 \times (9.4/\sqrt{10}) = 23.21$
Upper limit: $29.04 + 1.96 \times (9.4/\sqrt{10}) = 34.87$

Obviously, the interval estimates provide useful information for system risk evaluation. For instance, we can vary the repair time of this generator between 22 and 36 hours in sensitivity studies.

3.3.3 Confidence Interval of Variance

The following statistics theorem [16] provides the theoretical basis for estimation of the confidence interval of variance.

Theorem 3: If s^2 is the sample variance of a random sample of size n from the normal population $N(X; \mu, \sigma^2)$, then $(n-1)s^2/\sigma^2$ follows the chi-square distribution with $n-1$ degrees of freedom.

According to this theorem, for the given significance level α, it can be affirmed that the random variable $(n-1)s^2/\sigma^2$ is located between $\chi^2_{1-\alpha/2}(n-1)$ and $\chi^2_{\alpha/2}(n-1)$ with probability of $1-\alpha$, where $\chi^2_{\alpha/2}(n-1)$ is such a value that the integral of the χ^2 distribution density function from $\chi^2_{\alpha/2}(n-1)$ to ∞ equals $\alpha/2$. This concept is shown in Figure 3.3. Therefore, we have

$$\chi^2_{1-\alpha/2}(n-1) \leq \frac{(n-1)s^2}{\sigma^2} \leq \chi^2_{\alpha/2}(n-1) \tag{3.15}$$

or equivalently,

$$\frac{(n-1)s^2}{\chi^2_{\alpha/2}(n-1)} \leq \sigma^2 \leq \frac{(n-1)s^2}{\chi^2_{1-\alpha/2}(n-1)} \tag{3.16}$$

The values of $\chi^2_{\alpha/2}(n-1)$ and $\chi^2_{1-\alpha/2}(n-1)$ in Equation (3.16) can be obtained by looking up the χ^2 distribution table (see Appendix E).

3.4 ESTIMATING FAILURE FREQUENCY OF INDIVIDUAL COMPONENTS

As mentioned earlier, most data collection systems provide failure frequencies rather than failure rates. The sample mean method discussed above can be used to estimate the total average failure frequency based on the failure frequencies of mul-

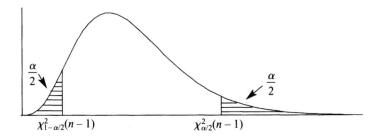

Figure 3.3 Confidence interval of variance using χ^2 distribution.

tiple components but cannot be applied to the estimation of the failure frequency for a single component.

3.4.1 Point Estimation

The failure frequency of an individual component is estimated as average failures per year over a time period:

$$f = \frac{N_f}{T} \tag{3.17}$$

where N_f is the number of repairable failures of a component in the time period T considered (in years). It is important to appreciate that if the forced failure frequency is calculated, T should be the elapsed time minus the shutdown time due to all nonforced outages (maintenance and other planned outages).

For transmission lines (overhead lines or cables), the average failure frequency of several lines should not be calculated using the sum of their individual failure frequencies divided by the number of lines, except for the failure frequency in per kilometer or 100 km, as in Table 3.2. This is because even if the lines are classified as the same type and carry the same voltage level, their lengths are generally different. In this case, the failure frequencies of all the lines should be normalized using the line lengths. Equation (3.18) is suggested for each individual line:

$$f_i = \frac{L_i \sum_{i=1}^{m} \frac{N_{fi}}{T_i}}{\sum_{i=1}^{m} L_i} \tag{3.18}$$

where f_i is the estimated failure frequency of the ith line, L_i the length (in kilometers) of the ith line, m the number of lines, and N_{fi} the number of failures of the ith line in the time period T_i. Note that if the forced failure frequency is calculated, T_i is varied for different lines since the time due to their nonforced outages, which needs to be excluded from the total elapsed time, is generally different. This equation becomes Equation (3.17) if one single line is considered. It should be stressed that it has been implicitly assumed in Equation (3.18) that the failure frequency of lines is proportional to the length. This assumption is generally reasonable. Otherwise, instead of using the total average, the individual failure frequency of each line should be used.

3.4.2 Interval Estimation

It is difficult to develop a theoretically accurate method for the interval estimation of failure frequency. Fortunately, statistics theory enables us to estimate the confidence interval of the failure rate, which can be approximately thought as an interval estimate of failure frequency, as they are numerically close in most cases.

In failure data processing, there exists the following relationship between the χ^2 distribution and the Poisson distribution [15]:

$$\chi^2(2F) = 2\lambda T \tag{3.19}$$

where λ is the average failure rate, T the total time length considered, and F the statistical number of failures within the time T in data samples.

This equation indicates that the quantity of two times the expected number of failures in the time T follows the χ^2 distribution with $2F$ degrees of freedom. Therefore, for the given significance level α, it can be asserted that the failure rate λ falls into the following random confidence interval with the probability of $1 - \alpha$:

$$\frac{\chi^2_{1-\alpha/2}(2F)}{2T} \leq \lambda \leq \frac{\chi^2_{\alpha/2}(2F)}{2T} \tag{3.20}$$

A more conservative estimate for the upper limit is

$$\frac{\chi^2_{1-\alpha/2}(2F)}{2T} \leq \lambda \leq \frac{\chi^2_{\alpha/2}(2F + 2)}{2T} \tag{3.21}$$

The following example is used to illustrate the concept. A 230 kV line had four forced failures in the past 10.5 years and the cumulative time for maintenance outages was 0.5 years. The average repair time was 100 hours. Estimate the confidence interval of the forced failure frequency for the given significance level of 0.05.

The total duration considered for calculating the forced failure frequency is $T = 10.5 - 0.5 = 10$ years. The average forced failure frequency is estimated as $4/10 = 0.4$ failures/year. Using Equation (2.7), the average failure rate is $0.4/(1 - 0.4 \times 100/8760) = 0.402$ failures/year. Obviously, the confidence interval for the failure rate can be used as that for the failure frequency since their point estimates are numerically so close.

With $2F = 2 \times 4 = 8$ and $\alpha/2 = 0.025$, we can obtain the following values by looking at Table 3 in Appendix E:

$$\chi^2_{1-\alpha/2}(2F) = 2.180 \quad \text{and} \quad \chi^2_{\alpha/2}(2F) = 17.535$$

Lower limit: $2.18/(2 \times 10) = 0.109$ failures/year
Upper limit: $17.535/(2 \times 10) = 0.877$ failures/year

If the conservative estimation formula is used, $\chi^2_{\alpha/2}(2F + 2) = 20.483$. Upper limit: $20.483/(2 \times 10) = 1.024$ failures/year.

A χ^2 distribution table is generally available for looking up the values of $\chi^2_{1-\alpha/2}(2F)$ and $\chi^2_{\alpha/2}(2F)$ with $2F$ degrees of freedom up to 30. When the degree of freedom is larger than 30, the following formula can be approximately used:

$$\chi^2 = \tfrac{1}{2}[\sqrt{(2\nu - 1)} + z]^2 \tag{3.22}$$

where ν is the degree of freedom and z is such a value that the integral of the standard normal-distribution density function from z to ∞ equals to the given significance level α. As a matter of fact, the values of the χ^2 distribution with any degree of freedom can be easily obtained using Microsoft EXCEL.

3.5 ESTIMATING PROBABILITY FROM A BINOMIAL DISTRIBUTION

In power system risk studies, it is often necessary to use the probability from a binomial distribution as input data, such as the probability of a generator's failure to start or the probability of a breaker's failure to switch. As a point estimate, such a failure probability can be easily estimated by

$$p = \frac{x}{n} \tag{3.23}$$

where x is the number of failure events and n the total number of actions (the number of starts or switching actions) in historical data records. When n is sufficiently large, p approaches its real probability.

Similarly, it is necessary to develop an interval estimation approach for the probability from the binomial distribution. As is well known from statistics theory [16], for a large n, the binomial population can be approximated using the normal distribution. It can be therefore concluded that the expression $(x - np)/\sqrt{np(1 - p)}$ can be approximately treated as if it were a random variable following the standard normal distribution, where x is the variable from the binomial distribution (failure events), p the failure probability, and n the number of experiments. Therefore, for the given significance level α, we have [16]

$$-z_{\alpha/2} \le \frac{x - np}{\sqrt{np(1 - p)}} \le z_{\alpha/2} \tag{3.24}$$

where $z_{\alpha/2}$ is such a value that the integral of the standard normal distribution from $z_{\alpha/2}$ to ∞ equals to $\alpha/2$. By solving this double inequality for p, the lower and upper limits of p with the confidence degree of $1 - \alpha$ are obtained as follows:

$$\frac{x + \frac{1}{2}z_{\alpha/2}^2 \pm z_{\alpha/2} \sqrt{\frac{x(n - x)}{n} + \frac{1}{4}z_{\alpha/2}^2}}{n + z_{\alpha/2}^2} \tag{3.25}$$

Expression (3.25) is an accurate solution of Inequality (3.24). A simpler approximation expression for a large sample can be developed. By rewriting Inequality (3.24) as

$$\frac{x}{n} - z_{\alpha/2}\sqrt{p(1 - p)/n} \le p \le \frac{x}{n} + z_{\alpha/2}\sqrt{p(1 - p)/n} \tag{3.26}$$

and substituting x/n for p in the two bounds we have

$$\frac{x}{n} - z_{\alpha/2}\sqrt{\frac{x}{n}\left(1 - \frac{x}{n}\right)/n} \leq p \leq \frac{x}{n} + z_{\alpha/2}\sqrt{\frac{x}{n}\left(1 - \frac{x}{n}\right)/n} \qquad (3.27)$$

A numerical example is given for illustration. The historical records show that a diesel generator failed to start 25 times in a total of 400 starts. Calculate the confidence interval of the probability of failure to start for the significance level of 0.05.

$$z_{0.025} = 1.96 \qquad x = 25 \qquad n = 400$$

The point estimate of the probability is $p = 25/400 = 0.0625$. Using Expression (3.25):

Lower limit: $(26.9208 - 9.6813)/403.8416 = 0.0427$
Upper limit: $(26.9208 + 9.6813)/403.8416 = 0.0906$

Using Inequality (3.27):

Lower limit: $0.0625 - 0.0237 = 0.0388$
Upper limit: $0.0625 + 0.0237 = 0.0862$

3.6 EXPERIMENTAL DISTRIBUTION OF FAILURE DATA AND ITS TEST

Two special issues are discussed in this section. The first one is how to create the experimental discrete probability distribution of a parameter from failure data records and the second one is how to use a statistical method to test whether the experimental distribution matches a theoretical distribution. When a Monte Carlo technique is used in system risk evaluation, it can simulate either a discrete or a theoretical distribution of any parameter in the component-outage models.

3.6.1 Experimental Distribution of Failure Data

In most cases, the average values of the parameters in the component-outage models (failure frequency, repair time, unavailability, etc.) are used as the input data in system risk evaluation. The point estimation of the parameters creates these values. In order to conduct sensitivity studies, a confidence range of the parameter is needed. The interval estimation provides a reference for the range. A more sophisticated way to handle the uncertainty of failure data is to consider their probability distributions. This basically means that any parameter can be of multiple values, with each having a probability.

A theoretical probability distribution is often used. In some cases, however, the theoretical distribution may not be valid and an experimental distribution has to be employed. Also, the experimental distribution serves to justify an assumed theoretical distribution. The procedure to produce the experimental distribution based on historical statistics of a parameter includes the following steps:

1. Rearrange the historical data records of a parameter in order of increasing value and define the total interval using the maximum and minimum values of the data.
2. Divide the total interval into k equal subintervals with the dividing points:

$$a_0 < \ldots < a_i < \ldots < a_k$$

3. Calculate the experimental probability of the data in each subinterval, which is the number of the data n_i ($i = 1, 2, \ldots, k$) falling into each subinterval $[a_{i-1}, a_i)$ divided by the total number of data n:

$$f_i = \frac{n_i}{n} \qquad (i = 1, 2, \ldots, k) \tag{3.28}$$

Obviously, $0 \le f_i \le 1$ and

$$\sum_{i=1}^{k} f_i = 1$$

When n is large enough, f_i approaches the real probability of the parameter in the subinterval $[a_{i-1}, a_i)$.

4. Build a function $f(x)$ by making its value in the subinterval $[a_{i-1}, a_i)$ be $f_i /(a_i - a_{i-1})$. The function $f(x)$ is the discrete probability density function of the experimental distribution and its graphical expression, as shown in Figure 3.4, is often called the sample histogram. It is important to appreciate that the discrete density function itself is random and must be updated as the number of failure data increases.

3.6.2 Test of Experimental Distribution

Although an experimental distribution can be easily simulated using Monte Carlo techniques, it is sometimes preferable to determine a theoretical distribution of a parameter based on the behavior of failure data. The theoretical distribution has a mathematical expression and is useful in developing the analytical formulas for risk assessments, which may simplify the calculation process in some cases. In principle, the theoretical distribution can be determined with the estimated mean and variance. For the distributions that do not explicitly contain the mean and standard deviation as the parameters (such as the Weibull or log-normal distribution), there is always the definite relationship between the parameters of the distribution function and their mean and standard deviation. This enables us to calculate the parameters from the estimated mean and variance.

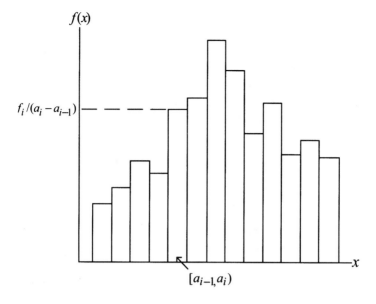

Figure 3.4   Sample histogram of an experimental distribution.

Before a theoretical distribution can be assumed, one necessary step is to assure no significant difference between the experimental and theoretical distributions through a statistical test. The χ^2 distribution is often used to conduct the statistical test. The significance of the test is twofold. On the one hand, if we know that a parameter follows a theoretical distribution, the purpose of the test is to accept or reject a set of data for the parameter. On the other hand, if we do not know the theoretical distribution of a parameter and want to assume one using the estimated mean and standard deviation, the purpose is to confirm or refuse the assumption.

Take the theoretical normal population as an example. Using the sample mean $\overline{X}$ and the sample standard deviation s to replace the mean μ and the standard deviation σ, the theoretical probability p_i of the normal population in each subinterval $[a_{i-1}, a_i)$ can be calculated by

$$P_i = \int_{(a_{i-1}-\bar{x})/s}^{(a_i-\bar{x})/s} \frac{1}{\sqrt{2\pi}} e^{(-t^2/2)} dt \tag{3.29}$$

The following statistical quantity is obtained using the experimental probability f_i and the theoretical probability p_i:

$$\chi^2 = n \sum_{i=1}^{k} \frac{(f_i - p_i)^2}{p_i} \tag{3.30}$$

This statistical quantity approximately follows the $\chi^2(k-3)$ distribution, where k is the number of pairs of f_i and p_i. Given the significance level α, if the χ^2 calculated using Equation (3.30) satisfies $\chi^2 \leq \chi^2_\alpha(k-3)$, the test is accepted, indicating no significant difference between the experimental and theoretical distributions. Otherwise, the test is rejected.

For other theoretical distributions, the test procedure is similar. In general, the degree of freedom of the χ^2 distribution is $k - q - 1$, where q is the number of the estimated distribution parameters. For example, $q = 2$ for the normal population ($\overline{X}$ and s), whereas $q = 1$ for the exponential population (λ).

3.7 ESTIMATING PARAMETERS IN AGING FAILURE MODELS

In order to calculate the unavailability due to aging failures using the methods given in Section 2.2.2, it is necessary to estimate the mean life of components and its standard deviation for the normal distribution model or the shape and scale parameters for the Weibull distribution model. Intuitively, it looks as if the mean life is nothing else than the average age of dead or retired components and that the sample mean concept could be used. Statistics theory also provides the maximum likelihood estimation approach. For the normal distribution, the maximum likelihood estimation still creates the sample mean as the estimated result, that is, the average age of dead components. For the Weibull distribution, it leads to an extremely complex set of equations with multiple solutions for the shape and scale parameters.

Unfortunately, the sample mean method is generally not workable for the mean life estimation of power system components, although it works well in the case where there is a large population, such as the average failure frequency and repair time for repairable failures. The power system components (generators, transformers, reactors, cables, etc.) have a long life, up to or even beyond 50 years, and, therefore, there are very limited aging failure data in a utility. It is possible to use the end-of-life failure records for the same type of equipment across utilities. However, the end-of-life failure data may be still insufficient, and the operational conditions, even for the same type of equipment, may vary greatly between different utilities that are geographically spread out and have different operation strategies and maintenance rules. In other words, the mean life, even for the same type of equipment, in different utilities may be quite varied.

When applied to the mean life estimation, the essential weakness of the sample mean is that only the information of dead components is used. For an equipment group with very few dead members, not only dead components but also survivors should make contributions to the mean life estimation. This is the basic idea in developing the mean life estimation method presented in this section. The concept can be interpreted using a simple example.

Let us imagine two equipment groups with the same total number of components. Assume that they have the same number of dead components and each one has died at the same age. If the survivors in the first group are older compared to those in the second one, it obviously implies that the first group should have a longer mean life than the second one. The estimates of the mean life and standard deviation obtained using this idea are the dynamic random variables and should be updated when more information on the survivors and dead components in an equipment group becomes available as the time advances. The variation in the estimate is not preferable but unavoidable in any estimation method. The purpose is to develop a relatively accurate mean life estimation method to deal with the situation with limited dead components.

3.7.1 Mean Life and Its Standard Deviation in the Normal Model

The presented method to estimate the mean life and standard deviation under the normal distribution model assumption includes the following steps:

Step 1: Collect the data of in-service years and retired years of components in an equipment group.

Step 2: Calculate the ages of each component. For a retired (dead) component, its age is the difference between the retired year and the in-service year. For a surviving component, its age is the difference between the current year and the in-service year.

Step 3: Calculate the numbers of exposed components and retired components for each age year in increasing order from age year 1, 2, . . . , to N, where N corresponds to the current year. Obviously, the smaller age year corresponds to the larger number of exposed components. Generally, only a few age years have retired component(s) and there is no retired component for the majority of age years.

Step 4: Calculate the discrete failure density probability of the equipment group and then the cumulative failure probability corresponding to each age year. The failure density probability is the number of retired components divided by the number of exposed components.

Step 5: Create a table that contains the following three columns:

First column: the age year

Second column: the cumulative failure probability of the equipment group that corresponds to the age year

Third column: the X-coordinate value z from the origin that corresponds to the area of the cumulative failure probability under the standard normal distribution density function curve. Given a cumulative probability, the X-coordinate value can be obtained by the inverse operation of the integral of the normal density function or by looking in the normal distribution table (see Appendix E).

Note that the table includes only those age years at which the equipment group has a nonzero and different value of cumulative failure probability.

Step 6: Augment the table in Step 5 by adding the following one or two lines:

> If there is no aging failure at age year 1, insert the following line at the beginning of the table as the first line:

> First column: the age year is the age of the first retired component minus 1

> Second column: the cumulative failure probability is 0.001

> Third column: the X-coordinate value corresponds to the cumulative failure probability 0.001, which is –3.0905.

> This line indicates that the cumulative failure probability before the first aging failure is assumed to be a negligible value.

> If there is no aging failure in the last age year, (i.e., no aging failure in the current year), add the following line at the end of the table as the last line:

> First column: the age year is the largest age among all the surviving components

> Second column: the cumulative failure probability is the same as that corresponding to the age year of the last retied component

> Third column: the X-coordinate value is the same as that corresponding to the age year of the last retired component

> This added line provides the information indicating that the cumulative failure probability does not increase after the last aging failure until the current year.

Step 7: Apply the least squares method to estimate the mean life and standard deviation.

> Denoting the age years by x_i, the corresponding X-coordinate values from the origin by z_i, and the number of the data pairs of x_i and z_i in the augmented table by N, we have the following linear relationship:

$$x_i = \mu + \sigma z_i + e_i \qquad (i = 1, 2, \ldots, n) \qquad (3.31)$$

> where μ and σ are the mean life and standard deviation, respectively, and e_i is the error indicating how far away each pair of x_i and z_i deviates from the standard normal distribution.

> Construct the following L function that is the sum of the squares of all the errors:

$$L = \sum_{i=1}^{N}(x_i - \mu - z_i\sigma)^2 \qquad (3.32)$$

> μ and σ can be found by letting $\partial L/\partial \mu = 0$ and $\partial L/\partial \sigma = 0$. Since these are a set of linear simultaneous equations, the analytical solution is ob-

tained as follows:

$$\mu = \bar{x} - \sigma\bar{z} \tag{3.33}$$

$$\sigma = \frac{S_{zx}}{S_{zz}} \tag{3.34}$$

where

$$\bar{x} = \frac{1}{N} \sum_{i=1}^{N} x_i$$

$$\bar{z} = \frac{1}{N} \sum_{i=1}^{N} z_i$$

$$S_{zx} = \sum_{i=1}^{N} (z_i - \bar{z})(x_i - \bar{x})$$

$$S_{zz} = \sum_{i=1}^{N} (z_i - \bar{z})^2$$

3.7.2 Shape and Scale Parameters in the Weibull Model

The Weibull distribution is often used to model an aging-failure process. It is characterized by the shape (β) and scale (α) parameters. In other words, its mean and standard deviation are not explicitly contained in the probability distribution function. The basic ideas in developing the method to estimate β and α include:

- As in the method for the normal distribution model, the information on both dead and surviving components should make contributions to the estimation.
- The model is based on an estimation of the shape (β) and scale (α) parameters of the Weibull distribution, whereas the mean and standard deviation can be obtained from β and α.
- The method should avoid the resolution process of a set of nonlinear simultaneous equations, which is needed in using the maximum likelihood estimation approach and may lead to difficulties in the selection of initial values and during the iterations of the resolution.
- The mean and standard deviation from the normal distribution model are used as the initial estimate so that the parameters from the normal and Weibull models have consistent results.

The method using the Weibull distribution model assumption includes the following steps:

Step 1: Obtain the initial estimate of the mean and standard deviation using the normal distribution model as described earlier.

Step 2: Calculate the initial estimate of shape and scale parameters from the initial mean and standard deviation. The calculation method can be found in Appendix A.3.4.

Step 3: Prepare the data from historical records. The procedure is the same as that described for the normal distribution model (see Steps 1 to 4 in Section 3.7.1). In fact, this work has been done in Step 1.

Step 4: Create a table containing the following two columns:

First column: the age year

Second column: the survival probability (reliability) of the equipment group, which is 1.0 minus the cumulative failure probability

Similarly, the table includes only those age years at which the equipment group has a nonzero and different value of survival probability.

Step 5: Augment the table in Step 4 by adding the following one or two lines:

If there is no aging failure at age year 1, insert the following line at the beginning of the table as the first line:

First column: the age year is the age of the first aging-failure component minus 1

Second column: the survival probability is 1.0

This line indicates that the survival probability before the first aging failure is assumed to be 100%.

If there is no aging failure in the last age year, (i.e., no aging failure in the current year), add the following line at the end of the table as the last line:

First column: the age year is the largest age among all the surviving components

Second column: the survival probability is the same as that corresponding to the age year of the last retired component

This added line provides the information indicating that the survival probability is kept unchanged after the last aging failure until the current year.

Step 6: Apply an optimization method to estimate the shape (β) and scale (α) parameters. The following survival function can be derived from the Weibull density function:

$$R = \exp\left[-\left(\frac{t}{\alpha}\right)^{\beta}\right] \tag{3.35}$$

where R is the survival probability and t is the age year. With the data in the table in Step 5, we have the following nonlinear relationship:

$$\ln R_i = -\left(\frac{t_i}{\alpha}\right)^{\beta} + e_i \qquad (i = 1, \ldots, N) \tag{3.36}$$

where e_i is the error between the data and the Weibull distribution and N is the number of the data in the table.

Similarly, construct the following L function that is the sum of the squares of all the errors:

$$L = \sum_{i=1}^{N}\left[\ln R_i + \left(\frac{t_i}{\alpha}\right)^{\beta}\right]^2 \tag{3.37}$$

Theoretically, β and α can be obtained by solving $\partial L/\partial\beta = 0$ and $\partial L/\partial\alpha$. However, this is a set of complex nonlinear equations with multiple solutions. It is difficult to select an appropriate resolution algorithm and initial values.

The best estimates of β and α can be obtained when L reaches its minimum. Any optimization approach is suitable to conduct the minimization. There are only two unknown variables and even the gradient descent technique can quickly converge. The procedure is summarized as follows:

1. Calculate the value of the objective function L and its gradient $\mathbf{G}$ at the initial estimate of the shape and scale parameters obtained in Step 2.

2. Calculate the next estimate of the parameters by stepping towards the negative gradient direction, that is,

$$\mathbf{Y}_{k+1} = \mathbf{Y}_k - b_k\mathbf{G}_k \tag{3.38}$$

where $\mathbf{Y}$ is the vector representing the estimate of β and α, $\mathbf{G}$ is the gradient vector, and b is the step length. The subscript k indicates that this is an iteration process. The b is determined by means of a bisection technique to assure descending of L.

3. The iteration proceeds until the convergence is reached, that is, L cannot be further decreased.

Step 7: Calculate the mean life (μ) and standard deviation (σ) from the shape and scale parameters obtained in Step 6. It should be noted that the parameters β and α are already sufficient for building the aging-failure model using the Weibull distribution. In other words, we do not need μ and σ for this purpose. However, the mean and standard deviation can provide a clear picture about the aging status of an equipment group. The calculation formulas are given as follows:

$$\mu = \alpha\Gamma\left(1 + \frac{1}{\beta}\right) \tag{3.39}$$

$$\sigma^2 = \alpha^2\left[\Gamma\left(1 + \frac{2}{\beta}\right) - \Gamma^2\left(1 + \frac{1}{\beta}\right)\right] \tag{3.40}$$

where the gamma function $\Gamma(*)$ can be approximately calculated by
[79]

$$\Gamma(x) = \sqrt{2\pi}x^{(x-0.5)} e^{-x}\left(1 + \frac{1}{12x}\right) \tag{3.41}$$

3.7.3 Example

A 500 kV reactor group is used as an example to demonstrate the estimation of the parameters in the aging-failure models. The group contains 100 reactors with only four retired units in the 31 years from 1969 to 2000. 1969 is the year at which the first set of reactors was placed in service and 2000 is the reference year for the estimation. Table 3.4 shows the raw data. If the sample mean method is used, the mean life for this group is estimated as the average age of the four retired reactors, which is $[(1996 - 1970) + (1989 - 1970) + (1996 - 1969) + (1997 - 1969)]/4 = 25$ years. Obviously, this is an overpessimistic estimate since the effects of the surviving reactors have been ignored.

Table 3.5 presents the numbers of the exposed and retired reactors for each age year. Tables 3.6 and 3.7 show the intermediate data (the augmented tables in the procedures described above) used in the estimation for the normal and Weibull distribution models, respectively. The estimated parameters using the two models are given in Table 3.8. The following observations can be made:

- The estimates of the mean life and standard deviation obtained using the normal and Weibull models are quite close. The estimates of the shape and scale parameters are 7.3 and 41, respectively. This suggests that the shape of the Weibull distribution is close to that of the normal distribution in this case.

- The estimate of the mean life is much longer than the 25 years obtained using the sample mean method. By looking at the raw data in Table 3.4, the age of 35 reactors has reached or exceeded 30 years already. Apparently, the estimated mean life of 37 or 38 years obtained using the presented methods should be more reasonable.

With the estimates of the mean life and standard deviation, the aging failure probability of a reactor at any age can be evaluated using the models described in Section 2.2.2. Table 3.9 presents the unavailability due to the aging failure and the probability of transition to the aging failure for the four ages. Note that these are not the ages of the actual reactors in the group but the four typical ages. It can be seen that the unavailability due to the aging failure or the probability of transition to the aging failure has a different value when using the normal and Weibull distribution models. Which model is used in actual applications should be based on the investigation into specific equipment groups. This question is still a challenge.

Table 3.4   Raw data of 500 kV reactors

No.	In-service year	Retired year	No.	In-service year	Retired year
1	1979		51	1976	
2	1979		52	1976	
3	1979		53	1976	
4	1981		54	1970	
5	1981		55	1970	
6	1981		56	1970	
7	1985		57	1981	
8	1985		58	1981	
9	1985		59	1981	
10	1979		60	1983	
11	1979		61	1983	
12	1979		62	1983	
13	1969		63	1984	
14	1969		64	1984	
15	1969		65	1984	
16	1969		66	1983	
17	1969		67	1983	
18	1969		68	1983	
19	1970	1996	69	1984	
20	1996		70	1984	
21	1970		71	1984	
22	1970		72	1983	
23	1970	1989	73	1983	
24	1989		74	1983	
25	1970		75	1984	
26	1970		76	1984	
27	1970		77	1984	
28	1970		78	1978	
29	1970		79	1978	
30	1970		80	1978	
31	1970		81	1969	1996
32	1970		82	1996	
33	1976		83	1969	
34	1976		84	1969	
35	1976		85	1969	
36	1976		86	1969	
37	1976		87	1969	
38	1976		88	1969	
39	1976		89	1969	
40	1976		90	1969	
41	1976		91	1969	
42	1976		92	1969	
43	1976		93	1969	
44	1976		94	1969	

(*continued*)

Table 3.4   *Continued*

No.	In-service year	Retired year	No.	In-service year	Retired year
45	1976		95	1969	1997
46	1976		96	1997	
47	1976		97	1969	
48	1976		98	1969	
49	1976		99	1969	
50	1976		100	1969	

Table 3.5   Numbers of exposed and retired reactors

Age year	Exposed number	Retired number
0	100	0
1	100	0
2	100	0
3	100	0
4	99	0
5	97	0
6	97	0
7	97	0
8	97	0
9	97	0
10	97	0
11	97	0
12	96	0
13	96	0
14	96	0
15	96	0
16	93	0
17	84	0
18	75	0
19	75	1
20	68	0
21	68	0
22	62	0
23	59	0
24	59	0
25	38	0
26	38	1
27	37	1
28	36	1
29	35	0
30	35	0
31	22	0

Table 3.6 Data used in the estimator (normal distribution)

Age	Cumulative failure probability	z
18	0.00100	−3.09052
19	0.01433	−2.18848
26	0.04065	−1.74359
27	0.06768	−1.49362
28	0.09545	−1.30809
31	0.09545	−1.30809

Table 3.7 Data used in the estimator (Weibull distribution)

Age	Surviving probability
18	1.00000
19	0.98567
26	0.95935
27	0.93232
28	0.90455
31	0.90455

Table 3.8 Estimated parameters

Parameter	Normal model	Weibull model	Sample mean
Mean life (years)	37.628	38.363	25.0
Standard deviation (years)	6.896	6.293	
Shape parameter		7.341	
Scale parameter		40.950	

Table 3.9 Aging failure probabilities at four typical ages

No.	Age	Unavailability due to aging failure		Probability of transition to aging failure	
		Normal	Weibull	Normal	Weibull
1	20	0.00007	0.00102	0.00015	0.00215
2	30	0.01398	0.01315	0.02944	0.02713
3	38	0.08713	0.05728	0.17319	0.11540
4	45	0.18307	0.15594	0.34681	0.30113

3.8 CONCLUSIONS

This chapter discusses parameter estimation methods. The parameters in the component-outage models can be the failure frequency and repair time, failure and repair rates (or transition rates between the states for a multiple-state model), unavailability, mean life and its standard deviation, and so on. They are the input data in power system risk assessment and are estimated from historical failure records. The quality of the data is key in risk evaluation. As is well known, uncertainties or even errors in historical statistics are unavoidable. An essential task in the parameter estimation is to reduce the impacts of the uncertainties or errors and enhance the accuracy.

The parameter estimation methods can be divided into three levels: the point, interval, and distribution estimations. The point estimation provides a single value for a parameter, which is the average of data. The average parameter has been widely used in power system risk assessment for its simplicity in applications, although there is uncertainty about it. The sample mean is the basic method for the point estimation. The interval estimation creates a confidence range of a parameter, which is extremely useful in sensitivity studies. This aids the investigation into the impacts of input data uncertainties on the results of system risk evaluation. There are different interval estimation approaches and they may produce slightly different confidence ranges. This does not weaken the significance of interval estimation. A pessimistic range can be used, depending on the necessity. Distribution estimation offers a probability distribution of a parameter. Using the probability distribution of input data can greatly enhance the accuracy of system risk evaluation. However, it needs more calculation efforts and higher requirements for assessment techniques.

There are two average concepts. The one is the average of sample data and the other is the average over a time period. The failure frequency of an individual component, which is associated with the second average concept, cannot be estimated using the sample mean method. The point and interval estimation methods for the individual component failure frequency have been discussed as a special topic. In addition to the general parameters like the failure frequency, repair time, and unavailability, the probability from the binomial distribution is also often used as the input data, such as the probability of a generator's failure to start or a breaker's failure to switch. The point and interval estimation methods for the probability following the binomial distribution have been presented.

The input data for aging failures are the mean life and standard deviation in the normal distribution model or the shape and scale parameters in the Weibull distribution model. The estimation methods for these parameters have been developed. The basic feature of the proposed methods is that the estimates can still be produced with an acceptable accuracy level even when there are very few end-of-life failure data, which is the usual situation for most utilities. In this case, the sample mean using the average age of retired components cannot provide a reasonable estimate of the mean life.

CHAPTER 4

ELEMENTS OF RISK
EVALUATION METHODS

4.1 INTRODUCTION

The outage models of individual components and the estimation methods of the parameters in the models have been discussed in Chapters 2 and 3. A system is a set of the components that are linked together and impact each other in some manner. System risk evaluation is used to assess the overall risk of the system due to individual component failures and their combinations. Many methodologies have been developed for engineering system risk evaluation [1–4]. The intent of this chapter is not to redescribe all existing methods but rather to review the most important concepts and general techniques that can be and have been used in power system risk assessment.

In Section 4.2, four fundamental methods are briefly discussed: the probability convolution, series and parallel networks, Markov equations, and frequency–duration approaches. The methods are selected from many general reliability techniques because they are not only conceptually important but also can be directly applied to simple cases in power system risk evaluation.

For a large-scale and complex power system, risk assessment requires two key processes. The first one is selection of system states and calculations of state probabilities. The second one is the problem analysis of selected system states and the remedial action taken for problems. There are two main methods for system state selection: state enumeration and Monte Carlo simulation. As mentioned in Chapter 1, both have merits and demerits. In general, if failure probabilities of components are quite small, the enumeration method is more efficient. When the number of severe events is relatively large and/or complex operating conditions are involved, the Monte Carlo method is often preferable. The basic concepts and procedures of state enumeration and Monte Carlo simulation methods are outlined in Section 4.3. The analysis techniques and remedial actions for system problems vary for different functional zones of power systems and will be discussed in Chapter 5.

The concepts and methods presented in this chapter are the theoretical basis of risk evaluation techniques for power systems.

4.2 METHODS FOR SIMPLE SYSTEMS

The four methods that are discussed in this section are the general methodologies for engineering system risk evaluation and are not limited to application in power systems. The main focus of the methods is on the index calculation. Power system risk assessment is much beyond this scope and includes other tasks such as failure criteria, different outage modes, system analyses, and remedial corrections. However, these methods provide the fundamental concepts of risk evaluation. As seen in Chapter 2, the Markov equation and the series network formula have been used in building the outage models of individual components. Other risk evaluation methods such as the fault tree, minimum cut-set, and event tree are not selected for discussion. The concept of the "or" and "and" gates in the fault tree is essentially consistent with the definition of series and parallel networks. Although the minimum cut-set was once addressed to assess the substation configuration risk, its application is actually very limited since it cannot model the breaker switching and dependent outages that have to be included in the substation risk evaluation. The principle used in the event tree is basically similar to the enumeration technique that is described in Section 4.3.1. The reader can refer to other textbooks for more materials on the general methods that are not covered in this chapter [1–8].

4.2.1 Probability Convolution

In some cases, a risk evaluation is nothing more than the probability convolution. For example, the essence of generation system risk evaluation is to calculate the difference between the generation capacity and the load demand as two random variables following probability distributions. Mathematically, this is the concept of convolution.

The mathematical definition of convolution is as follows. Assume that two independent random variables X and Y have the probability density functions $g(x)$ and $h(y)$, respectively. The probability density function $s(z)$ of the random variable $Z = X - Y$ is the following integral, which is called the convolution:

$$s(z) = \int_{-\infty}^{+\infty} g(w) \cdot h(w - z)dw \qquad (4.1)$$

In power system risk evaluation, a risk index is often the mean value of the random variable Z. Theoretically, the mean can be calculated with the probability density function $s(z)$. The definition in Equation (4.1) is accurate but not suitable for use in actual applications. It is better to have the discrete expression of the difference between the two random variables since random variables in power system risk evaluation are often discrete.

Given the two random variables X and Y with the following discrete probability density functions,

$$p(X = X_i) = p_i \qquad (i = 1, \dots, n)$$
$$p(Y = Y_k) = p_k \qquad (k = 1, \dots, m)$$

(4.2)

the mean value of the random variable $Z = X - Y$ is calculated by

$$\bar{Z} = \sum_{i=1}^{n} \sum_{k=1}^{m} (X_i - Y_k) p_i p_k$$

(4.3)

It should be appreciated that a risk index is usually the mean value of Z under a given condition. For instance, if X represents the load demand and Y the generation capacity, the expected demand not supplied should be the mean of Z under the condition of X larger than Y. It is not difficult to incorporate the condition in the discrete convolution. Equation (4.4) includes this condition:

$$\bar{Z} = \sum_{i=1}^{n} \sum_{k=1}^{m} \max(0, X_i - Y_k) p_i p_k$$

(4.4)

Even if X and Y are continuous random variables, their probability density functions can be easily discretized to use Equation (4.4). This avoids the continuous integration required by Equation (4.1).

4.2.2 Series and Parallel Networks [2]

The series or parallel network is defined here as the logical relationship between the failure (or success) of the network and the failures (or successes) of its components.

Series network: The components are said to be in series if only one needs to fail for the network to fail or they must be all up for network success.

Parallel network: The components are said to be in parallel if they must all fail for the network to fail or only one needs to be up for network success.

Obviously, the series or parallel network does not refer to a topological structure of power system components, although it is often consistent with the topology of a physical system. The failure manner of a power system structure must be analyzed to determine if the series and parallel network concept can be used to model the system or not. A radial distribution system generally can be modeled using the series network concept. A simple substation configuration can be represented using a network composed of series and parallel branches.

The definitions of series and parallel networks above are essentially the same as the union and intersection concepts in probability theory and, therefore, the equations developed below can have a wider application beyond physical system structures. For example, the total equivalent parameters of the component-related and terminal-related failures described in Section 2.2.7 are a specific application of the series concept.

The following derivation is based on a two-component network and the resulting equations can be easily extended to multiple components.

4.2.2.1 Series Network. Consider the case of two repairable components in series, as shown in Figure 4.1. From the definition of the series network, the following relationships hold:

$$U_{se} = U_1 + U_2 - U_1U_2 \tag{4.5}$$

$$\lambda_{se} = \lambda_1 + \lambda_2 \tag{4.6}$$

$$A_{se} = A_1A_2 \tag{4.7}$$

Here, A denotes the availability, U the unavailability, and λ the failure rate. The r, μ, and f in the following are the repair time, repair rate, and failure frequency. The subscripts 1, 2, and se represent Components 1, 2, and the equivalent series network, respectively. Equation (4.7) can be rewritten as

$$(1 - U_{se}) = (1 - U_1)((1 - U_2) \tag{4.7A}$$

For a repairable system, applying Equations (2.1) and (4.7A), we obtain

$$\left(1 - \frac{\lambda_{se}}{\lambda_{se} + \mu_{se}}\right) = \left(1 - \frac{\lambda_1}{\lambda_1 + \mu_1}\right)\left(1 - \frac{\lambda_2}{\lambda_2 + \mu_2}\right) \tag{4.8}$$

By substituting Equation (4.6) into (4.8) and applying the relationship in Equation (2.3), the equivalent repair time for the series network can be calculated by

$$r_{se} = \frac{\lambda_1 r_1 + \lambda_2 r_2 + \lambda_1 r_1 \lambda_2 r_2}{\lambda_1 + \lambda_2} \tag{4.9}$$

Applying the relationships in Equations (2.5), (2.7), (4.6), and (4.7A) gives

$$f_{se} = f_1(1 - f_2 r_2) + f_2(1 - f_1 r_1) \tag{4.10}$$

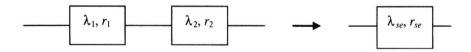

Figure 4.1   Series network equivalence.

Equations (4.5) to (4.7), (4.9), and (4.10) are used to calculate the risk indices of the series network. Equations (4.5) to (4.7) apply to both repairable and nonrepairable failures, whereas the others only to repairable failures. Although these formulas are derived from the network of two components, it is straightforward to apply them to the case of multiple components in series. The idea is to use them for any two components first. Then the network equivalence of the first two components is treated as a component and combined with the third one, and so on. The repeated use of the equations is an easy process in computer programming or on an EXCEL spreadsheet.

4.2.2.2 Parallel Network. Figure 4.2 shows the case of two repairable components in parallel. From the definition of the parallel network, the following relationships hold:

$$U_{pe} = U_1 U_2 \tag{4.11}$$

$$\mu_{pe} = \mu_1 + \mu_2 \tag{4.12}$$

$$A_{pe} = A_1 + A_2 - A_1 A_2 \tag{4.13}$$

All the symbols are the same as those defined in Section 4.2.2.1. Applying (2.3) and (4.12) yields

$$r_{pe} = \frac{r_1 r_2}{r_1 + r_2} \tag{4.14}$$

For repairable components, using Equations (2.1) and (4.11) gives

$$\frac{\lambda_{pe}}{\lambda_{pe} + \mu_{pe}} = \left(\frac{\lambda_1}{\lambda_1 + \mu_1}\right)\left(\frac{\lambda_2}{\lambda_2 + \mu_2}\right) \tag{4.15}$$

Considering the relationship in Equation (2.3) and substituting Equation (4.14) into (4.15), we obtain

$$\lambda_{pe} = \frac{\lambda_1 \lambda_2 (r_1 + r_2)}{1 + \lambda_1 r_1 + \lambda_2 r_2} \tag{4.16}$$

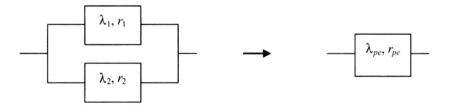

Figure 4.2   Parallel network equivalence.

Applying Equations (4.14), (4.16) and the relationship in Equation (2.6) produces

$$f_{pe} = f_1 f_2 (r_1 + r_2) \qquad (4.17)$$

Equations (4.11) to (4.14), (4.16), and (4.17) are used to calculate the risk indices for the parallel network. Equations (4.11) and (4.13) apply to both repairable and nonrepairable failures, whereas the others only to repairable failures. Similarly, these formulas can be repeatedly employed when multiple components are considered. A network consisting of both series and parallel branches can be evaluated through combined uses of the equations given in Sections 4.2.2.1 and 4.2.2.2.

4.2.3   Markov Equations [2]

The Markov equation approach is sometimes called the state space method since it is based on a state space diagram. The main advantage of this technique is the clear picture of all states and transitions between them. It is extremely useful in modeling the outages of individual components, as shown in Chapter 2. One demerit is the difficulty in applying it to a large system. For a system containing N components with each having two states (up and down), the number of system states is 2^N. When N is relatively large, it is almost impossible to draw a state space diagram.

The Markov method can be used to solve both the time dependent and limiting state probabilities. The former is associated with a set of differential equations and the latter with a set of algebraic equations. The power system risk evaluation is normally a limiting state probability problem and only the Markov equation for limiting state probabilities is discussed.

The system of two repairable components is used as an example to explain the Markov method. The procedure includes the following steps:

Step 1: Construct a state space diagram according to the transitions of component states. Figure 4.3 shows the four states and their transitions for the system of two repairable components.

Step 2: Build the transition matrix based on the state space diagram. The dimension of the matrix is the same as the number of system states, that is, each state corresponds to one row and one column. Check the system states in the state space diagram one by one. If there is a transition from State i to State j ($i \neq j$), the transition rate is filled as the element at the ith row and jth column. Otherwise, the element is filled by zero. The diagonal element in each row is 1.0 minus the sum of all the other elements in this row. For the given example, the matrix appears as

$$T = \begin{array}{c} \\ 1 \\ 2 \\ 3 \\ 4 \end{array}\begin{array}{c} \\ \left[\begin{array}{cccc} 1 - (\lambda_1 + \lambda_2) & \lambda_1 & \lambda_2 & 0 \\ \mu_1 & 1 - (\mu_1 + \lambda_2) & 0 & \lambda_2 \\ \mu_2 & 0 & 1 - (\mu_2 + \lambda_1) & \lambda_1 \\ 0 & \mu_2 & \mu_1 & 1 - (\mu_1 + \mu_2) \end{array}\right] \end{array} \qquad (4.18)$$

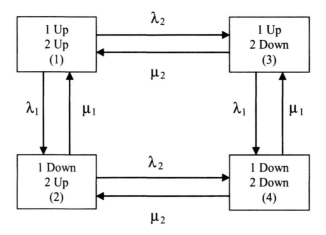

Figure 4.3 State space diagram of two repairable components.

It should be noted that the resulting Equation (4.18) is not a probability matrix since λ and μ are not probabilities. It is just a way to develop the Markov equation for solving the limiting state probabilities.

Step 3: Apply the Markov approach, which states that the limiting state probabilities would not change in the further transition process. Mathematically, it can be expressed as

$$PT = P \tag{4.19}$$

where P is the limiting state probability vector and T the transition matrix. Equation (4.19) can be rewritten as

$$P(T - I) = 0 \tag{4.20}$$

where I is the unit matrix. For the given example, Equation (4.20) has the following full matrix form:

$$[P_1 \ P_2 \ P_3 \ P_4] \begin{bmatrix} -(\lambda_1 + \lambda_2) & \lambda_1 & \lambda_2 & 0 \\ \mu_1 & -(\mu_1 + \lambda_2) & 0 & \lambda_2 \\ \mu_2 & 0 & -(\mu_2 + \lambda_1) & \lambda_1 \\ 0 & \mu_2 & \mu_1 & -(\mu_1 + \mu_2) \end{bmatrix}$$
$$= [0 \ 0 \ 0 \ 0] \tag{4.21}$$

Transposing Equation (4.21) gives the algebraic matrix equation in the general form

$$
\begin{bmatrix}
-(\lambda_1 + \lambda_2) & \mu_1 & \mu_2 & 0 \\
\lambda_1 & -(\mu_1 + \lambda_2) & 0 & \mu_2 \\
\lambda_2 & 0 & -(\mu_2 + \lambda_1) & \mu_1 \\
0 & \lambda_2 & \lambda_1 & -(\mu_1 + \mu_2)
\end{bmatrix}
\begin{bmatrix}
P_1 \\ P_2 \\ P_3 \\ P_4
\end{bmatrix}
=
\begin{bmatrix}
0 \\ 0 \\ 0 \\ 0
\end{bmatrix}
\tag{4.22}
$$

Step 4: Add the full probability condition—the sum of the probabilities of all system states should be 1. For the given example, that is

$$
[P_1 + P_2 + P_3 + P_4] = 1.0 \tag{4.23}
$$

It is important to appreciate that the Markov matrix equation obtained in Step 3 has a rank of $N - 1$, where N is the number of system states. In other words, only $N - 1$ equations are independent and the full probability condition has to be added. In the given example, any one in the four equations in (4.22) must be replaced by Equation (4.23) in order to solve the four system state probabilities. For instance, the first equation in (4.22) is replaced by Equation (4.23). This gives

$$
\begin{bmatrix}
1 & 1 & 1 & 1 \\
\lambda_1 & -(\mu_1 + \lambda_2) & 0 & \mu_2 \\
\lambda_2 & 0 & -(\mu_2 + \lambda_1) & \mu_1 \\
0 & \lambda_2 & \lambda_1 & -(\mu_1 + \mu_2)
\end{bmatrix}
\begin{bmatrix}
P_1 \\ P_2 \\ P_3 \\ P_4
\end{bmatrix}
=
\begin{bmatrix}
1 \\ 0 \\ 0 \\ 0
\end{bmatrix}
\tag{4.24}
$$

Step 5: Solve the Markov matrix equation obtained in Step 4 using a linear algebraic algorithm. For the given example, the solution is as follows:

$$
P_1 = \frac{\mu_1 \mu_2}{(\mu_1 + \lambda_1)(\mu_2 + \lambda_2)} \tag{4.25}
$$

$$
P_2 = \frac{\lambda_1 \mu_2}{(\mu_1 + \lambda_1)(\mu_2 + \lambda_2)} \tag{4.26}
$$

$$
P_3 = \frac{\mu_1 \lambda_2}{(\mu_1 + \lambda_1)(\mu_2 + \lambda_2)} \tag{4.27}
$$

$$
P_4 = \frac{\lambda_1 \lambda_2}{(\mu_1 + \lambda_1)(\mu_2 + \lambda_2)} \tag{4.28}
$$

Step 6: If necessary, the frequency and duration can be calculated using the frequency-duration approach described in the next subsection.

4.2.4 Frequency-Duration Approaches [2]

The frequency-duration approach is basically a technique to calculate the frequency and duration from the state probabilities and transition rates. Although the frequency balance concept can also be used to construct the state space equations for limiting state probabilities, this can be viewed as a derivative of the Markov method and is not discussed here.

4.2.4.1 Frequency of Encountering a State.
The frequency of encountering State i is calculated by

$$f_i = P_i \sum_{k=1}^{M_d} \lambda_k = \sum_{j=1}^{M_e} P_j \lambda_j \qquad (4.29)$$

where P_i is the probability of State i, P_j is the probability of a state directly communicating to State i, λ_k or λ_j is the transition (failure or repair) rate, M_d is the number of the rates departing from State i, and M_e is the number of the rates entering State i.

Equation (4.29) also indicates the basic concept in the frequency balance approach, that is, the frequency of leaving a state is equal to the frequency of entering the state for any state in an ergodic system. For example, the frequency of leaving or encountering State 1 in Figure 4.3 is

$$f_1 = P_1(\lambda_1 + \lambda_2) = P_2\mu_1 + P_3\mu_2 = \frac{\mu_1\mu_2(\lambda_1 + \lambda_2)}{(\mu_1 + \lambda_1)(\mu_2 + \lambda_2)} \qquad (4.30)$$

4.2.4.2 Frequency of Transition between Two States.
The frequency of transition from State i to State j is calculated by

$$f_{ij} = P_i \lambda_{i-j} \qquad (4.31)$$

where P_i is the probability of State i and λ_{i-j} the transition rate from State i to State j. Generally, $f_{ij} = f_{ji}$ if there exist both the transitions from State i to State j and from State j to State i in an ergodic system. For example, the two transition frequencies between States 1 and 2 in Figure 4.3 are

$$f_{12} = f_{21} = P_1\lambda_1 = P_2\mu_1 = \frac{\mu_1\mu_2\lambda_1}{(\mu_1 + \lambda_1)(\mu_2 + \lambda_2)} \qquad (4.32)$$

4.2.4.3 Frequency of Encountering a State Set.
In power system risk evaluation, system states are grouped into sets in terms of their outcome. For instance, all the states leading to load curtailments are aggregated to give a system failure state set.

The probability of residing in a state set is simply the sum of the probabilities of all the states in the set. The frequency of encountering a state set is evaluated by

$$f_s = \sum_{k \in s} f_k - \sum_{i,j \in s} f_{ij} \qquad (4.33)$$

where f_s is the frequency of encountering the state set S, f_k the frequency of encountering State k, which belongs to S, and f_{ij} the frequency from State i to State j. The subscripts i and j denote the two states having a direct communication within the set S. It should be noted that both f_{ij} and f_{ji} should be included in the second term.

For example, consider States 3 and 4 in Figure 4.3 as a set. The frequency of encountering the state set is

$$\begin{aligned}
f_{s(34)} &= f_3 + f_4 - f_{34} - f_{43} \\
&= P_3(\mu_2 + \lambda_1) + P_4(\mu_2 + \mu_1) - P_3\lambda_1 - P_4\mu_1 \qquad (4.34) \\
&= (P_3 + P_4)\mu_2
\end{aligned}$$

It can be seen from the state space diagram in Figure 4.3 that the essential implication of Equation (4.33) is to treat the state set as a superstate and only those transitions cross the boundary between the superstate and other states are considered.

4.2.4.4 Mean Duration of Residing in Each System State.
The mean duration of residing in a state is the reciprocal of the sum of all departure rates and can be calculated directly from the state space diagram:

$$d_i = \frac{1}{\displaystyle\sum_{k=1}^{M_d} \lambda_k} \qquad (4.35)$$

where d_i is the mean duration of residing in State i. λ_k and M_d are the same as defined in Equation (4.29).

For State 1 in Figure 4.3, for instance, the mean duration is

$$d_1 = \frac{1}{\lambda_1 + \lambda_2} \qquad (4.36)$$

4.2.4.5 Mean Duration of Residing in a State Set.
Substituting Equation (4.35) into (4.29) gives

$$P_i = f_i d_i \qquad (4.37)$$

This equation indicates that the state probability can be factored as the product of frequency and duration. This is the most important concept in the frequency-duration approach.

The relationship between P, f, and d in Equation (4.37) is general. In other words, it applies not only to a single state but also to a state set.

If Equation (4.37) is applied to State 1 in Figure 4.3, the same result as Equation (4.36) is obtained:

$$d_1 = \frac{P_1}{f_1} = \frac{\mu_1\mu_2}{(\mu_1 + \lambda_1)(\mu_2 + \lambda_2)} \cdot \frac{(\mu_1 + \lambda_1)(\mu_2 + \lambda_2)}{\mu_1\mu_2(\lambda_1 + \lambda_2)} = \frac{1}{(\lambda_1 + \lambda_2)} \qquad (4.38)$$

Applying Equation (4.37) to the set of States 3 and 4 gives

$$d_{s(34)} = \frac{P_3 + P_4}{f_{s(34)}} = \frac{P_3 + P_4}{(P_3 + P_4)\mu_2} = \frac{1}{\mu_2} \tag{4.39}$$

4.3 METHODS FOR COMPLEX SYSTEMS

The methods discussed in Section 4.2 are conceptually important and can also be applied in some simple cases. However, it is difficult to directly apply those methods to a large-scale transmission system or a complex substation configuration, although they may be used to model a portion of the complex systems. This section introduces the more sophisticated methods that are used in power system risk evaluation: state enumeration and Monte Carlo simulation. The latter can be categorized into sequential and nonsequential sampling. The basic idea in the methods for a complex system is to perform risk evaluation using an iteration process including the following four steps:

1. Selecting a system state
2. Analyzing the system state to judge if it is a failure state
3. Calculating risk indices for the failure state
4. Updating cumulative indices

The state enumeration and Monte Carlo simulation are essentially the two different methods for selecting system states. The formulas for calculating indices have different expressions in the two methods. However, the system analysis does not rely on how to select a system state and is the same for both the methods. Generation, distribution, substation, and transmission systems need different system analysis techniques. Generally, the techniques are relatively simple and straightforward for generation and distribution systems, more complicated for substation configurations, and most complex for transmission networks. The system analysis techniques for the different functional zones of power systems along with the details of state selection will be discussed in Chapter 5. The focus of this section is on the basic concepts of state enumeration and Monte Carlo simulation. It is important to recognize that any method has both advantages and disadvantages. There is no all-purpose method. The merits and demerits of each method will be commented on. Selection of a method depends on its appropriateness to a particular system problem, data requirements, and complexity of evaluation.

4.3.1 State Enumeration

State enumeration is based on the expansion of the following expression:

$$(P_1 + Q_1)(P_2 + Q_2) \cdots (P_N + Q_N) \tag{4.40}$$

where P_i and Q_i are the success and failure probabilities of the ith component and N is the number of components in the system.

The probability of a system state is given by

$$P(s) = \prod_{i=1}^{N_f} Q_i \prod_{i=1}^{N-N_f} P_i \qquad (4.41)$$

where N_f and $N - N_f$ are the numbers of failed and nonfailed components in State s respectively. For the normal state in which all the components are up, $N_f = 0$ and the equation becomes

$$P(s) = \prod_{i-1}^{N} P_i \qquad (4.42)$$

According to the concepts of frequency and mean duration of a system state given in Section 4.2.4, the system state frequency and the mean duration can be calculated using Equations (4.43) and (4.44):

$$f(s) = P(s) \sum_{k=1}^{N} \lambda_k \qquad (4.43)$$

$$d(s) = \frac{1}{\sum_{k=1}^{N} \lambda_k} \qquad (4.44)$$

Here, λ_k is the departure rate of the kth component in State s. If the kth component is up, λ_k is the failure rate and if is down, λ_k is the repair rate.

It can be seen from Equation (4.41) that all the enumerated system states are mutually exclusive. Therefore, the cumulative system failure probability is the direct sum of the probabilities of all failure states:

$$P_f = \sum_{s \in G} P(s) \qquad (4.45)$$

where G is the set of all system failure states.

According to Equation (4.33), the cumulative system failure frequency is given by

$$F_f = \sum_{s \in G} f(s) - \sum_{n,m \in G} f_{nm} \qquad (4.46)$$

where f_{nm} denotes the transition frequency from State n to State m. The second term implies that all the transition frequencies between system failure states must be deducted from the sum of the frequencies of all system failure states. This is a very difficult task in the state enumeration technique, if not impossible. In actual power system risk evaluation, the deduction is often neglected. This leads to the following approximate formula for the cumulative system failure frequency:

$$F_f = \sum_{s \in G} f(s) \tag{4.47}$$

The approximation is generally acceptable since the transitions between system failure states are very rare, whereas the transitions between the normal and failure system states are dominant in real life.

Once the cumulative system failure frequency is obtained, the mean duration of residing in the set of system failure states is calculated by

$$D_f = \frac{P_f}{F_f} \tag{4.48}$$

It should be noted that the mean failure duration is also approximate if F_f is calculated by Equation (4.47).

For each system failure state, any other risk index function, such as load curtailment, $C(s)$, can be obtained through system analysis techniques. The mathematical expectation of the index function for all system failure states is given by

$$E(C) = \sum_{s \in G} C(s)P(s) \tag{4.49}$$

Note that nonfailure system states have a zero value of the index function and do not make contributions to $E(C)$, although their probabilities are not zero.

It is important to recognize the following points in actual applications:

- A necessary condition of the state enumeration method is the mutual exclusiveness between all enumerated system states. Some literatures and even some commercial programs only consider the unavailability of failed components in calculating the probability of a system failure state. This is mathematically incorrect. When the number of components is large, and particularly when components have relatively low availability, neglecting the second portion of Equation (4.41) may lead to a large error.

- It is not computationally feasible to enumerate all system states for a system containing a large number of components since the number of system states increases exponentially with the number of components. A common practice is to stop at a given enumeration depth, which is often denoted by a failure level. The first failure level refers to the system states containing only one component failure, the second failure level to those containing two component failures, and so on. An alternative criterion is to specify a sufficiently small threshold of system state probability. The system states whose probability is lower than the threshold are ignored.

- If both repairable and aging failures of components are considered, the Q_i in the equation can be the total unavailability that is the union of the two unavailability values due to repairable and aging failures while the P_i is ($1.0 - Q_i$). In this case, no calculation associated with the frequency and duration is needed or can be performed since there is no concept of frequency and repair time for aging (end-of-life) failures.

- Compared to the Monte Carlo methods, state enumeration is more effective for a system with a relatively small number of components and/or low failure probabilities of components.

- It should be kept in mind that the frequency index obtained using Equation (4.47) is an upper-bound estimate of the real frequency as the transition frequencies between failure system states have not been excluded. The relevant mean duration should be a lower-bound estimate.

- It should be remembered that chronologically time-dependent events could not be modeled in the state enumeration method. Such events may or may not be necessary to consider in a particular case. This depends on the user's judgment and understanding of a system.

4.3.2 Nonsequential Monte Carlo Simulation [1, 13, 14]

Nonsequential Monte Carlo simulation is sometimes called the state sampling approach. It is widely used in power system risk evaluation. The concept is based on the fact that a system state is a combination of all component states and each component state can be determined by sampling the probability of the component appearing in that state.

Each component can be modeled using a uniform distribution between [0, 1]. Assume that each component has two states of failure and success and component failures are independent of each other. Let s_i denote the state of the ith component and Q_i its failure probability. Produce a random number R_i distributed uniformly between [0, 1] for the ith component:

$$s_i = \begin{cases} 0 & \text{(success)} & \text{if } R_i > Q_i \\ 1 & \text{(failure)} & \text{if } 0 \leq R_i \leq Q_i \end{cases} \tag{4.50}$$

The state of the system containing N components is expressed by the vector s:

$$s = (s_1, \ldots, s_i, \ldots, s_N) \tag{4.51}$$

After a system state is selected in the sampling, the system analysis is performed to judge whether it is a failure state or not, and if yes, a risk index function for that state is evaluated.

When the number of samples is sufficiently large, the sampling frequency of the system state s can be used as an unbiased estimate of its probability:

$$P(s) = \frac{m(s)}{M} \tag{4.52}$$

where M is the number of samples and $m(s)$ is the number of occurrences of the system state s in the sampling.

Once the probability of each system state is estimated in the sampling, the same formulas as those in Equations (4.43) to (4.49) can be applied to calculate the system failure probability, system failure frequency, mean system failure duration, and other system risk indices.

It can be seen that the main difference between nonsequential simulation and state enumeration is how to select system states and how to calculate probabilities of individual system states.

More details of the nonsequential Monte Carlo simulation method can be found in Reference [1]. The following aspects should be appreciated in an actual application.

- A necessary step is generation of a random number sequence for each component. The random numbers must meet the three basic requirements: uniformity, independence, and long repeat cycle.

- The Monte Carlo simulation is a fluctuating process. Therefore, the estimated risk indices always come with a confidence band. There is no guarantee that a few more samples will definitely lead to a smaller error. It is true, however, that the confidence band decreases as the number of samples increases.

- An appropriate convergence criterion is key to assure accuracy in a Monte Carlo simulation. The coefficient of variance is often used as the stopping rule in the sampling. In power system risk evaluation, different risk indices have varied convergence speeds. It has been found that the coefficient of variance of the EENS index (expected energy not supplies) has the lowest rate of convergence and therefore should be used as the convergence criterion in a multiple index study. An alternative is to use a prespecified maximum number of samples as the stopping rule. When the simulation process ends, the coefficient of variance is checked to see whether it is small enough. If not, a new run with an increased number of samples is necessary. This alternate approach is used when the user has no idea on how much CPU time is needed to reach a sufficiently small coefficient of variance.

- The nonsequential simulation only needs failure probabilities of components as input data in the sampling process. This feature enables us to easily simulate the unavailability due to both repairable and aging failures. Two independent random numbers are created for a component, one for the unavailability of repairable failures and another for the unavailability due to aging failures.

- The idea of state sampling not only applies to component-failure events but also can be generalized to sample the states for other parameters in power system risk evaluation, such as load levels, hydrological and weather states, and so on. Also, this method is not limited to simulations on a yearly basis but can be easily used to perform simulations over any time length (weekly, monthly, seasonal, or yearly).

- Compared to the state enumeration, the state sampling method is a preferable option when a relatively large system or a system with relatively high failure probabilities of components is evaluated. In such cases, the state enumeration method may require much more CPU time to achieve the same accuracy level.

- Similar to state enumeration, nonsequential Monte Carlo simulation cannot capture the chronology of time-dependent events so that the system failure frequency and mean failure duration are approximately estimated.

4.3.3 Sequential Monte Carlo Simulation [1]

Sequential Monte Carlo method refers to a simulation process over a chronological time span. There are different approaches to create an artificial system state transition cycle. The most popular one is the so-called state duration sampling, which is discussed here. Another approach—system state transition sampling—can be found in Reference [40].

The state duration sampling approach is based on sampling a probability distribution of component state duration and includes the following steps:

Step 1: Specify initial states of all components. Generally, all components are assumed to be in the up state initially.

Step 2: Sample the duration of each component residing in its present state. The probability distribution of the state duration should be assumed. Different states such as operation or repair processes may assume varied probability distributions for the state duration. For example, the sampling value of the state duration following an exponential distribution is given by

$$D_i = \frac{1}{\lambda_i} \ln R_i \qquad (4.53)$$

where R_i is a uniformly distributed random number between [0, 1] corresponding to the ith component. If the present state is the up state, λ_i is the failure rate of the ith component, and if the present state is the down state, λ_i is its repair rate. Appendix B.4 provides several approaches to generate the random variates following different probability distributions.

Step 3: Repeat Step 2 in the time span considered (years) and record sampling values of each state duration for all components. The chronological state transition processes of each component in the given time span can be obtained as shown in Figure 4.4.

Step 4: Create the chronological system state transition cycle by combining the state transition processes of all components. This concept is shown in Figure 4.5.

Step 5: Conduct the system analysis for each different system state to calculate the risk index functions. The calculations for the system risk indices are

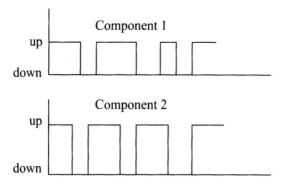

Figure 4.4   Chronological state transition processes of components.

straightforward since the occurrences of system failure states and their duration and consequences can be clearly identified and recorded in the system transition cycle. Equations (4.54)–(4.56) are the three general formulas of the risk indices:

$$P_f = \frac{\displaystyle\sum_{k=1}^{M_{dn}} D_{dk}}{\displaystyle\sum_{k=1}^{M_{dn}} D_{dk} + \sum_{j=1}^{M_{up}} D_{uj}} \tag{4.54}$$

$$F_f = \frac{M_{dn}}{\displaystyle\sum_{k=1}^{M_{dn}} D_{dk} + \sum_{j=1}^{M_{up}} D_{uj}} \tag{4.55}$$

$$D_f = \frac{\displaystyle\sum_{k=1}^{M_{dn}} D_{dk}}{M_{dn}} \tag{4.56}$$

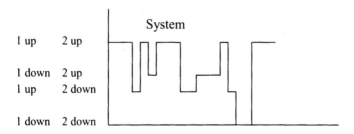

Figure 4.5   Chronological system state transition process.

Here, P_f, F_f, and D_f are the system failure probability, frequency, and mean duration, respectively. D_{dk} is the duration in the kth down state and D_{uj} is the duration in the jth up state. M_{dn} and M_{up} are, respectively, the numbers of occurrences of system failure and success states in the simulation span. These two numbers are generally the same, except for the case where a failure or success state is truncated at the end of the sampling span.

It can be seen that key to the sequential Monte Carlo method is generation of a system state transition process. Once this is completed, the index calculations are simple. The essence of the method is to create an artificial cycle of system operations and failures.

It is important to recognize the following points in an actual application:

- One crucial step in the sequential Monte Carlo method is to calculate sampling values of the state duration random variable following a probability distribution. The basis is generation of a random number uniformly distributed between [0, 1] (see Appendix B.2 and B.4).

- As in the nonsequential Monte Carlo method, the sequential simulation is also a fluctuating process and, therefore, an appropriate convergence criterion is needed. The coefficient of variance can be still used as the stopping rule. However, it should be noted that the number of samples in the sequential method is not the number of system states but the number of sampling years.

- The main advantages of the sequential Monte Carlo method are accurate evaluation of frequency and duration indices, flexibility of modeling any state-duration distribution, and the capacity of calculating statistical probability distributions of system risk indices. These are the weaknesses of the state enumeration or nonsequential simulation method.

- Compared to the nonsequential Monte Carlo simulation, the sequential Monte Carlo method requires more CPU time and storage space. Besides, it requires the parameters associated with all component state-duration distributions. Even for the exponential distribution assumption, these are the transition rates between all possible states of each component. In some cases, especially for a multiple state component representation, it may be difficult to obtain all the input data required.

- The sequential simulation is based on the chronological concept and cannot be used to conduct simulations for a case with a nonchronology feature. For instance, if the time frame of a study is limited to a month, say, September, it would be incorrect to simulate the case over a series of Septembers, as a September is not chronologically followed by another September.

- It is impossible to simulate the unavailability model of aging failures given in Section 2.2.2 using the sequential Monte Carlo method. This is because an aging failure has been assumed to be end-of-life and does not have the concept of failure frequency and repair time, whereas the sequential simulation is based on a transition process containing many failures and repairs. However, the sequential Monte Carlo method can be used to simulate a process with alternate aging failures and replacements. It must be stressed that the latter case is completely different from the unavailability model of aging failures, in which only aging failures but not replacements are considered. A replacement is conceptually different from a repair.

4.4 CONCLUSIONS

This chapter reviews the basic concepts and fundamental methods that can be and have been used in power system risk evaluation. The concepts and methods are general and can also be used in the risk assessment of other engineering systems.

The four methods selected for simple systems are the probability convolution, series and parallel networks, Markov equations, and frequency-duration approaches. As will be seen in the next chapter, a discrete probability convolution can be directly applied in generation system risk evaluation. The series and parallel networks of repairable components can be used for the risk assessment of simple distribution systems and substation configurations. The concepts of the series and parallel networks are consistent with the union and intersection in probability theory. This enables them to have wider applications. Simply speaking, any problem that can be modeled by the union and intersection concepts can be represented using series and parallel networks. The Markov equation and its state space diagram provide a means to clearly express the transition relationship between states. In addition to the use in building outage models of system components, it can be also applied in some system cases with a limited number of components. The frequency-duration approach is mainly utilized in the cases where frequency and duration indices need to be evaluated.

Two primary methods are presented for the risk assessment of large-scale systems: state enumeration and Monte Carlo simulation. The Monte Carlo simulation is further divided into sequential and nonsequential sampling. The basic procedures of the methods have been discussed. Particularly, the key points for each of the methods that should be appreciated in actual applications, including the merits and demerits, are commented on. These comments are useful information for selecting the method appropriate to a specific system or problem. A power system generally has a large size so that the two methods are the main tools used in power system risk evaluation. The most important point to be recognized is that power system risk assessment requires not only selection of system states and calculations of state probabilities, but also the consequence analysis of selected system states and the remedial correction of failed states. In implementation, risk evaluation techniques for power systems are much more complex than the basic concepts that are addressed in this chapter. The details of the risk evaluation techniques for the different functional zones of power systems will be discussed in Chapter 5.

CHAPTER 5

RISK EVALUATION TECHNIQUES FOR POWER SYSTEMS

5.1 INTRODUCTION

Power systems are divided into the functional zones of generation, transmission, substation, and distribution. Generation system risk assessment is concerned with only generation facilities and is called the hierarchical level one (HL1) study. The hierarchical level 2 (HL2) includes both the generation and transmission equipment, whereas the hierarchical level 3 (HL3) comprises all the functional zones. In actual applications, HL2 studies are sometimes performed only for transmission system risk evaluation by assuming that the generation is 100% reliable when a transmission network is the main object under investigation. Although substation configurations are thought of as a portion of a transmission system, substation risk assessment is generally carried out separately because the failure criteria and evaluation techniques for transmission systems and substation configurations are different. However, it is possible to incorporate substation configurations into a transmission network risk assessment. This case will be specifically discussed in Chapter 9. Usually, HL3 studies are not directly conducted due to enormity of the problem. Instead, distribution system risk is assessed as a separate entity. HL3 indices can be evaluated by using bus indices from the HL2 study as input data in the distribution risk analysis.

This chapter illustrates the risk evaluation techniques for generation, distribution, substation, and composite generation and transmission systems, respectively, in Sections 5.2 to 5.5. For each functional zone, both analytical and Monte Carlo simulation methods are discussed, system analysis techniques are presented, and formulas of risk indices are derived.

Risk Assessment of Power Systems. By Wenyuan Li
ISBN 0-471-63168-X © 2005 the Institute of Electrical and Electronics Engineers, Inc.

5.2 TECHNIQUES USED IN GENERATION-DEMAND SYSTEMS

A generation-demand system is usually called a generation system. The risk evaluation of a generation system provides the indices measuring overall adequacy but not at a single substation or customer load point since the network between generations and loads has been omitted. The generation-demand system model is shown in Figure 5.1. It represents generators on one side and a total load demand on other side. In other words, the model deals with two random variables of generation and load. Both have multiple MW levels, with each level having a probability of occurrence. The system analysis logic is straightforward. For a system state with generator failures, if the total load is larger than the total generation, a load curtailment is needed to maintain the power balance. The task of generation system evaluation is to quantify the risk due to generator's random failures. Combinations of load curtailments and their probabilities for all possible system states create a system failure risk index.

5.2.1 Convolution Technique

Mathematically, generation system risk assessment is to conduct a convolution of the two random variables of generation and load. The analytical convolution technique includes the following three steps:

1. Creating a discrete generation probability distribution
2. Creating a discrete load probability distribution
3. Performing a convolution between the two distributions

5.2.1.1 Discrete Generation Probability Distribution. The discrete generation probability distribution can be obtained using the enumeration method given by Equations (4.40) and (4.41). The concept is illustrated using a simple numerical example as follows.

A system consists of two 10 MW and one 15 MW generating units with the identical unavailability of 0.03. Using Equation (4.40), we have

$$
\begin{array}{ccc}
10\,\text{MW} & 10\,\text{MW} & 15\,\text{MW} \\
(0.97 + 0.03) & (0.97 + 0.03) & (0.97 + 0.03)
\end{array}
$$

All the generation capacity levels with their probabilities are enumerated and shown in Table 5.1. Note that the states of the same MW level have been merged with a cumulated probability in the table. This is sometimes called the capacity outage probability table in the literature. In essence, it is a discrete generation probability distribution.

When the number of generators is large, the table becomes very lengthy since the number of generation levels increases exponentially. Two main measures are utilized to reduce the number of generation levels. First, a small probability value is used as a threshold and all the generation levels with a probability lower than the

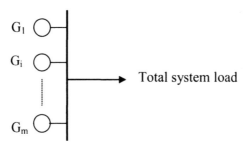

Figure 5.1 Generation-demand system model.

threshold are tailored from the table. Second, a reasonable interval of generation capacity is specified and the generation levels within the interval are rounded to the nearest level. The two measures produce some computational errors, but calculation efforts are reduced. This is a compromise between accuracy and CPU time.

5.2.1.2 Discrete Load Probability Distribution. A load duration curve must be considered in generation system risk evaluation. The multiple-level model shown in Figure 5.2 can be used to represent the original load curve [46]. The more the load level steps there are, the more accurate the model is. Given the load levels, a discrete load probability distribution can be obtained by assigning load points to the nearest level. This distribution, expressed as the load levels and their probabilities, is given in Table 5.2, where L_k is the kth load level, n the number of load levels, T the total time length of the load curve, and T_k the time length for the kth load level.

If the uncertainty of loads is considered, an augmented load level probability table can be created. Usually, a normal distribution is used to characterize the uncertainty of loads. The normal distribution can be modeled using several discrete intervals and each interval is represented by its midpoint. Take seven intervals as an example. This concept is shown in Figure 5.3 and Table 5.3. The "interval" column in the table indicates the locations of the midpoints of each interval using the standard deviation σ.

Table 5.1 Generation capacity and probability

Capacity (MW)		Probability
Available	Outage	
35	0	$0.97 \times 0.97 \times 0.97 = 0.912673$
25	10	$0.97 \times 0.97 \times 0.03 \times 2 = 0.056454$
20	15	$0.97 \times 0.97 \times 0.03 = 0.028227$
15	20	$0.97 \times 0.03 \times 0.03 = 0.000873$
10	25	$0.97 \times 0.03 \times 0.03 \times 2 = 0.001746$
0	35	$0.03 \times 0.03 \times 0.03 = 0.000027$

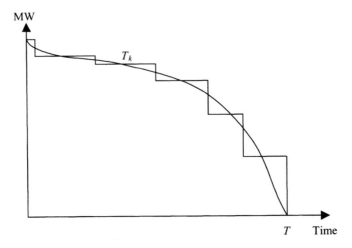

Figure 5.2   Load duration curve and its multiple-level step model.

By combining Table 5.2 with Table 5.3 using an enumeration concept, the augmented table of load levels and their probabilities is obtained (Table 5.4). This table indicates that each of the original load levels has been replaced by seven levels and the probability of each new load level is the probability of the original level weighted by one of the seven discrete probability values for the normal distribution. Note that σ_k here refers to the ratio of the standard deviation to the kth original load level. Each original load level may have a different standard deviation. Generally, however, the same standard deviation for all the original load levels is assumed in most actual applications.

5.2.1.3 Index Calculation. Once the generation capacity and load level probability tables are created, the convolution approach can be used to evaluate the risk

Table 5.2   Load levels and probabilities (uncertainty not included)

Load level	Probability
L_1	$P_1 = T_1/T$
L_2	$P_2 = T_2/T$
$\vdots$	$\vdots \quad \vdots$
L_k	$P_k = T_k/T$
$\vdots$	$\vdots \quad \vdots$
L_n	$P_n = T_n/T$

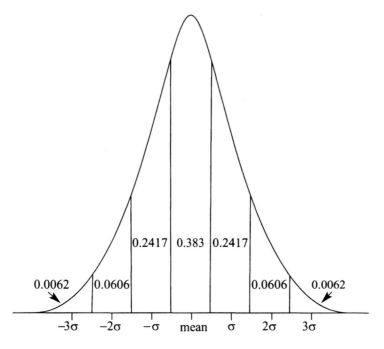

Figure 5.3 Seven-interval representation of normal distribution.

Table 5.3 Seven-interval representation of normal distribution

Interval	Probability
-3σ	0.0062
-2σ	0.0606
-1σ	0.2417
mean	0.3830
$+1\sigma$	0.2417
$+2\sigma$	0.0606
$+3\sigma$	0.0062

Table 5.4 Load levels and probabilities (uncertainty included)

Load level	Probability
⋮	⋮
$L_k(1 - 3\sigma_k)$	$P_{k1} = 0.0062T_k/T$
$L_k(1 - 2\sigma_k)$	$P_{k2} = 0.0606T_k/T$
$L_k(1 - 1\sigma_k)$	$P_{k3} = 0.2417T_k/T$
L_k	$P_{k4} = 0.3830T_k/T$
$L_k(1 + 1\sigma_k)$	$P_{k5} = 0.2417T_k/T$
$L_k(1 + 2\sigma_k)$	$P_{k6} = 0.0606T_k/T$
$L_k(1 + 3\sigma_k)$	$P_{k7} = 0.0062T_k/T$
⋮	⋮

indices of a generation system. The LOLE (loss of load expectation) and LOEE (loss of energy expectation) are given by Equations (5.1) to (5.3):

$$LOLE = \sum_{j=1}^{N_G} \sum_{i=1}^{N_L} P_i P_j I_{ij} \cdot T \tag{5.1}$$

$$I_{ij} = \begin{cases} 0 & L_i \le G_j \\ 1 & L_i > G_j \end{cases} \tag{5.2}$$

$$LOEE = \sum_{j=1}^{N_G} \sum_{i=1}^{N_L} P_i P_j \cdot \max(0, L_i - G_j) \cdot T \tag{5.3}$$

Here, L_i is the ith load level, P_i the probability of the ith load level, and N_L the number of the load levels in the load level probability table. G_j is the jth generation capacity level, P_j the probability of the jth generation capacity level, and N_G the number of the generation capacity levels in the generation capacity probability table. T is the total time length of the load duration curve. It should be noted that the units of LOLE and LOEE are hours per period and megawatt hours per period, respectively. The time period depends on the load curve considered and is often one year in the evaluation for system development planning. However, other time lengths such as one month or one season are also used in the evaluation for operation planning.

It should be appreciated that the convolution technique cannot be used to calculate frequency and duration indices. Although, theoretically, the frequency-duration approach described in Section 4.2.4 can be used for this purpose, it is not practical for a large generation system with many generators, particularly in the case including a load duration curve. In general, the state duration sampling method given in Section 5.2.3 is employed when the frequency and duration indices need to be evaluated.

5.2.2 State Sampling Method

The basic idea of the state sampling method for generation-demand system risk evaluation is to select the states of generators through the state sampling technique and still use a multiple-level step model for a load curve. The use of the analytical load level probabilities plays a role like the variance reduction technique in the nonsequential Monte Carlo sampling and produces better convergence performance. If the uncertainty of the load around each step level is considered, it can be modeled using a normally distributed random variable.

For each load level in the multiple step model shown in Figure 5.2, a uniformly distributed random number R_j is drawn between $[0, 1]$ for each generator. The state of the jth generator is determined by

$$s_j = \begin{cases} 0 \ (\text{up}) & \text{if } R_j > PP_j + PF_j \\ 1 \ (\text{down}) & \text{if } PP_j < R_j \le PP_j + PF_j \\ 2 \ (\text{derated}) & \text{if } 0 \le R_j \le PP_j \end{cases} \tag{5.4}$$

where PF_j is the probability in the down state (i.e., the unavailability) and PP_j the probability in the derated state. Obviously, this sampling concept can be easily extended to model multiple derated states of a generator without any increase in computational effort. This is the main advantage that makes the sampling method superior to the convolution technique.

The available capacity of each generator is determined according to its state so that the total system generation capacity can be obtained. For a given load level, the demand not supplied (*DNS*) in the *k*th sampling is calculated by

$$DNS_k = \max\left\{0, L_i - \sum_{j=1}^{m} G_{jk}\right\} \tag{5.5}$$

where L_i is the load at the *i*th level, G_{jk} the available capacity of the *j*th generator in the *k*th sampling and *m* the number of generators in the system.

In the case in which the uncertainty of the load is considered, the load level L_i is used as the mean, with the uncertainty represented by a standard deviation σ_i in a percentage of L_i. A standard normal distribution random number X_k is created using the approximate inverse transformation method given in Appendix B.4.2. The sampled value of the load in the *k*th sampling is given by

$$L_{\sigma i} = (X_k \sigma_i + 1)L_i \tag{5.6}$$

The $L_{\sigma i}$ is used to replace L_i in Equation (5.5) in order to capture the uncertainty of the load.

The risk indices are estimated using Equations (5.7) and (5.9):

$$LOLE = \sum_{i=1}^{N_L}\left(\frac{T_i}{N_i} \sum_{k=1}^{N_i} I_k(DNS_k)\right) \tag{5.7}$$

where I_k is the indicator variable, which means that

$$I_k(DNS_k) = \begin{cases} 0 & \text{if} & DNS_k = 0 \\ 1 & \text{if} & DNS_k \neq 0 \end{cases} \tag{5.8}$$

$$LOEE = \sum_{i=1}^{N_L}\left(\frac{T_i}{N_i} \sum_{k=1}^{N_i} DNS_k\right) \tag{5.9}$$

In the equations, N_L is the number of the load levels in the multiple step load model shown in Figure 5.2, T_i the time length of the *i*th load level, and N_i the number of samples at the *i*th load level. The units of LOLE and LOEE are hours per time period and megawatt hour per time period, respectively, where the time period can be a year, a season or a month.

In an actual application, the number of samples at each load level may or may not be the same. Generally, more samples are needed at a lower load level to reach the same accuracy. However, a lower load level always makes a smaller contribution to the indices and the lower accuracy is acceptable at very low load levels. Similar to the convolution technique, the state sampling method cannot provide the frequency and duration indices.

5.2.3 State Duration Sampling Method [1]

In the state duration sampling method, an artificial system generation capacity curve is created and superimposed on a chronological load curve to obtain a simulated operation history.

The first step is to generate the operating cycles of each generator by drawing sample values of time-to-failure and time-to-repair of generating units. The system generation capacity curve can be obtained by combining the operating cycles of all generators. This is shown in Figure 5.4.

The second step is to superimpose the system generation capacity curve on the chronological hourly load curve to obtain a system available margin model. A negative margin indicates that a system load has to be curtailed. The model gives the amount and duration of each load curtailment in a long simulation time span. Figure 5.5 shows the superimposition. If the uncertainty of loads needs to be considered, the load curve has to be modified. To obtain the modified load curve reflecting the load uncertainty, Equation (5.6) should be used to revise load values at each hour point on the load curve. The original hourly load is the mean and a standard deviation representing the uncertainty is applied. The modification process is associated with considerable calculation since the simulation time can be up to several thousands of years or even longer. However, the complete modification on all hourly points is not necessary. Each hour point on the superimposed model is checked. If the system available capacity at that point is larger than the original hourly load plus three times the standard deviation, this point is

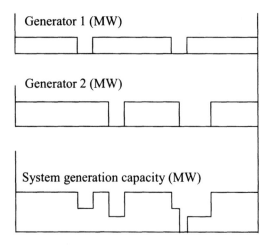

Figure 5.4 Generation capacity cycles of generators and system.

skipped as the probability that a sampled load value from the normal distribution sampling is larger than the system available capacity will be extremely low and can be ignored. The majority of load points on the load curve fall into this situation. Most load points that need to be modified using the normal distribution sampling should be those close to the load curtailment points before the modification (see the ENS locations in Figure 5.5).

The third step is to calculate the risk indices. In each sampling year i, the loss of load occurrence (LLO_i), loss of load duration (LLD_i) in hours, and energy not supplied (ENS_i) in megawatt hours can be directly obtained by observing the superimposed margin model. The risk indices in N sampling years can be estimated using Equations (5.10) to (5.13):

(1) Loss of load expectation (LOLE, h/yr):

$$LOLE = \frac{1}{N}\sum_{i=1}^{N} LLD_i \qquad (5.10)$$

(2) Loss of energy expectation (LOEE, MWh/yr):

$$LOEE = \frac{1}{N}\sum_{i=1}^{N} ENS_i \qquad (5.11)$$

(3) Loss of load frequency (LOLF, occurrences/yr):

$$LOLF = \frac{1}{N}\sum_{i=1}^{N} LLO_i \qquad (5.12)$$

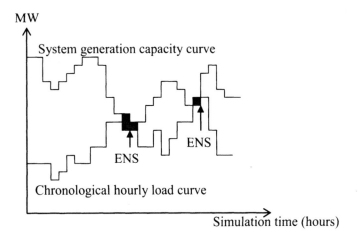

Figure 5.5 Superimposition of the system generation capacity curve on the hourly load curve.

(4) Loss of load duration (LOLD, hours/occurrence):

$$LOLD = \frac{\sum\limits_{i=1}^{N} LLD_i}{\sum\limits_{i=1}^{N} LLO_i} \qquad (5.13)$$

The main advantage of the state duration sampling method is that it provides not only the LOLE and LOEE indices but also the frequency and duration indices. The price paid for this is considerable CPU time in the simulation. Unless the frequency and duration indices are important or chronological time-dependent factors need to be considered, it is not desirable to use this method in most cases. This method cannot be used to consider a non-whole-year load curve.

It is worth stressing that the number of samples used in the state sampling method and the number of sampling years used in the state duration sampling are two totally different concepts. It is important to appreciate the difference.

5.3   TECHNIQUES USED IN RADIAL DISTRIBUTION SYSTEMS

Most distribution systems are designed to operate in a radial structure although some of them may have a few meshed loops. The risk evaluation techniques for radial configurations are presented in this section; the distribution systems with meshed loops can use the similar techniques for transmission systems that will be described in Section 5.5. Radial distribution systems have a set of series components between a substation and a load point. The components include the main feeder, lateral distributors, breakers, transformers, switches, and fuses. A failure of any of the components in series leads to the outage of the load point. The sectionalizing equipment provides a means of isolating a faulted section while maintaining as many healthy sections as possible to be supplied. In some systems, there is an alternate source to supply the sections that become disconnected from the main source after the faulted section has been isolated. Figure 5.6 shows a simple and typical radial distribution configuration.

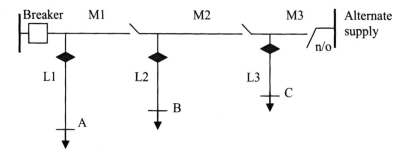

Figure 5.6   A sample radial distribution system.

5.3.1 Analytical Technique [3]

The analytical approach to the risk assessment of radial distribution systems has been widely used for years. It is easy to use in calculating the load point and system average performance indices. Since the components between the source (substation) and load points are essentially in series in a radial distribution system, the concept of the series network can be directly applied. Equation (4.9) gives the equivalent repair time for two components in series. Generally, the condition of $\lambda_1 r_1 \lambda_2 r_2 \ll \lambda_1 r_1$ or $\lambda_2 r_2$ holds in the distribution systems so that Equation (4.9) is approximated by

$$r_s = \frac{\lambda_1 r_1 + \lambda_2 r_2}{\lambda_1 + \lambda_2} \tag{5.14}$$

Equations (4.6) and (5.14) can be further extended to the case of any number of components in series:

$$\lambda_s = \sum \lambda_i \tag{5.15}$$

$$r_s = \frac{\sum \lambda_i r_i}{\sum \lambda_i} \tag{5.16}$$

Applying Equations (2.1) and (2.3) to the equivalent series network yields

$$U_s = \frac{\lambda_s}{\lambda_s + \dfrac{1}{r_s}} \approx \lambda_s r_s = \sum \lambda_i r_i \tag{5.17}$$

In the equations, λ, r, and U are the failure rate, repair time, and unavailability, respectively. The subscript i refers to Component i and the subscript s to the equivalent series network, which represents the path from the source to a load point. Note that the unavailability is often expressed in hours/year in distribution risk assessment, whereas it normally has no unit as a probability value at other occasions.

The sample system in Figure 5.6 is used to illustrate the application. The basic data are given in Table 5.5 and the calculation procedure is shown in Table 5.6. It should be noted that the recovery time of the power supply at a load point depends on the repair time if no switching action is performed, whereas it is reduced to the switching time if a switch or the alternate source can act. Introduction of switching actions decreases the duration of interruptions, leading to an improvement in reliability. In this example, a manual switching action is considered and does not have effects on improving the failure rates of load points. If an automatic switching device is used, a fault can be cleared almost right away. In such a case, if the fault event is so short that it can be classified as a nonfailure event, then the overall failure rate will be reduced to a value close to that caused only by lateral section failures.

Once the indices at load points are obtained, they can be used to calculate the system performance indices. For example, by assuming 250, 100, and 50 customers

Table 5.5   Data in the sample system

Feeder	Length (mile)	Failure rate (f/mile/yr)	Repair time (h/f)
M1	2	0.1	3.0
M2	3	0.1	3.0
M3	1	0.1	3.0
L1	3	0.25	1.0
L2	2	0.25	1.0
L3	1	0.25	1.0

Manual sectionalizing time of any switch: 0.5 h.
Switch-in time of the alternate source: 1.0 h.

Table 5.6   Calculation procedure

	Load point A			Load point B			Load point C		
Component	λ (f/yr)	r (h/f)	λr (h/yr)	λ (f/yr)	r (h/f)	λr (h/yr)	λ (f/yr)	r (h/f)	λr (h/yr)
Main section									
M1	0.2	3.0	0.6	0.2	1.0	0.2	0.2	1.0	0.2
M2	0.3	0.5	0.15	0.3	3.0	0.9	0.3	1.0	0.3
M3	0.1	0.5	0.05	0.1	0.5	0.05	0.1	3.0	0.3
Lateral section									
L1	0.75	1.0	0.75	—	—	—	—	—	—
L2	—	—	—	0.5	1.0	0.5	—	—	—
L3	—	—	—	—	—	—	0.25	1.0	0.25
Point indices	1.35	1.15	1.55	1.1	1.5	1.65	0.85	1.24	1.05

at the load points A, B, and C, respectively, the system performance indices are calculated as follows:

The total annual customer interruptions:

$$\sum N_s \lambda_s = (250)(1.35) + (100)(1.1) + (50)(0.85) = 490$$

The total annual customer interruption duration:

$$\sum N_s U_s = (250)(1.55) + (100)(1.65) + (50)(1.05) = 605$$

System average interruption frequency index (SAIFI):

SAIFI $= \sum N_s \lambda_s / \sum N_s = 490/400 = 1.225$ interruptions/system customer/year

System average interruption duration index (SAIDI):

SAIDI $= \sum N_s U_s / \sum N_s = 605/400 = 1.513$ hours/system customer/year

Customer average interruption duration index (CAIDI):

$$\text{CAIDI} = \sum N_s U_s / \sum N_s \lambda_s = 605/490 = 1.235 \text{ hours/customer interruption}$$

Average service availability index (ASAI):

$$\text{ASAI} = \left(\sum 8760 N_s - \sum N_s U_s \right) / \sum 8760 N_s = (400 \times 8760 - 605)/(400 \times 8760) = 0.999827$$

5.3.2 State Duration Sampling Method

The component state duration sampling method can also be applied in the risk assessment of radial distribution systems. The main purpose of using this method is to include the probability distributions of the failure rates and repair times of feeder sections and calculate the probability distributions of indices in the evaluation. The method includes the following steps:

Step 1: With the given failure rates and repair times of feeder sections, the artificial up–down–up operating cycles of all sections are generated using the state duration sampling approach described in Section 4.3.3. The failure rates and repair times can be assumed to follow any probability distribution.

Step 2: A system operating history is obtained by combining the up–down–up operating cycles of feeder sections and the up–down–up operating histories at all load points are created by examining the effects of section faults on each load point based on the system configuration and switching logic.

Step 3: The modifications associated with manual switching actions are conducted. For example, if the power supply at a load point can be restored through a switching action earlier than a repair, the down state duration of the load point caused by a particular feeder section failure should be adjusted to the switching time. It should be recognized that the modifications are generally varied for different load points because of different switching locations. The modifications are also varied at the different positions on the artificial operating history curves since sampled values of a repair time are different.

Step 4: The average outage rate, interruption duration and unavailability of load points are calculated using the following equations:

$$\lambda_i = \frac{M_i}{\sum T_{ui}} \tag{5.18}$$

$$r_i = \frac{\sum T_{di}}{M_i} \tag{5.19}$$

$$U_i = \frac{\sum T_{di}}{\sum T_{ui} + \sum T_{di}}$$

(5.20)

Here, λ_i, r_i, and U_i are, respectively, the average outage rate, interruption duration, and unavailability at the ith load point. $\sum T_{ui}$ and $\sum T_{di}$ are the total up and down times (years) at the ith load point in the sampling span. M_i is the total number of the transitions between the up and down states at the ith load point. Note that the unavailability U_i obtained using Equation (5.20) has no unit. If the unit of hours/year is used for U_i, it should be multiplied by 8760.

Step 5: The probability distribution of a load point outage rate can be evaluated by

$$P(k) = \frac{n(k)}{N} \qquad (k = 0, 1, 2, \ldots)$$

(5.21)

where N is the number of years in the whole sampling and $n(k)$ is the number of the years in which the load point outage rate equals k. Similar calculations can be performed to obtain the probability distributions of load point outage duration (r) and load point unavailability (U). Since r and U are the continuous quantities, discrete intervals must be specified.

More materials and details of the Monte Carlo simulation methods used in the risk evaluation of radial distribution systems can be found in Reference [1].

5.4   TECHNIQUES USED IN SUBSTATION CONFIGURATIONS

Substations provide the connections between generating units and a transmission system (voltage step-up substations) or between a transmission system and distribution systems (voltage step-down substations). The risk evaluation for voltage step-up and step-down substations is basically the same except that the former calculates the indices associated with loss of generations whereas the latter calculates those associated with loss of loads. The voltage step-down substation is used below as an example to discuss the risk assessment techniques for substation configurations. There are many types of substation arrangements, such as a single bus with or without sectional busbars, double or triple buses with or without sectional busbars, and ring busbars. The primary components in a substation include transformers, short cables, breakers, switches, and bus sections. Figure 5.7 shows a simple substation layout. The basic system analysis in the risk evaluation of substation configurations is mainly associated with failure modes of components and identification of the connectivity between sources and load points. The substation arrangements generally cannot be decomposed into the series and parallel structures except in some very simple cases.

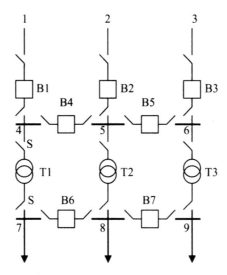

Figure 5.7 Substation configuration.

5.4.1 Failure Modes and Modeling

There are two basic failure modes for a substation component: passive and active. A passive failure refers to one that does not cause any operation of protection devices and therefore does not have any impact on remaining healthy components. Service is restored after repairing the failed component. An open circuit is an example of a passive failure. An active failure is one that causes an operation of the primary protection zone around the failed component and therefore causes the outage of other healthy components. The actively failed component is isolated by switching actions, leading to service being restored to some or all of the load points. The failed component itself enters a repairing state and is restored only after the repair. A short fault is an example of an active failure. For a breaker or switch, a stuck condition is the third failure mode. This failure mode creates a wider protection zone and the outage of more healthy components. The maintenance outage is also often considered in substation risk assessment.

The typical four-state model for a substation component is given in Figure 5.8. This model includes forced passive failure, forced active failure, and maintenance outage, but not stuck breaker conditions. As pointed out in Section 2.2.4, a maintenance outage can be also modeled using a scheduled event if the Monte Carlo simulation method is used. In this case, the state space model becomes a three-state model, whereas the scheduled maintenance is considered through modifying the simulated operating history. It should be noted that the switching state is caused by an active failure and thus is associated with multiple component outages. Take the short circuit fault on T1 in Figure 5.7 as an example. The fault will cause Breakers B1, B4, and B6 to be opened. In other words, T1 and these three breakers should all be out of service in the connectivity analysis for the switching state.

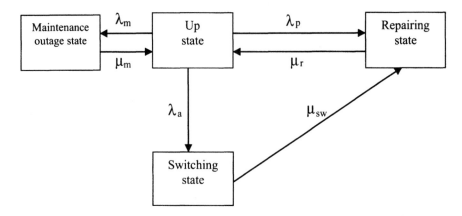

Figure 5.8 Four-state model of a substation component.

If stuck breakers are considered, more states must be introduced into the state space diagram. For instance, for the short-circuit fault on T1 in Figure 5.7, the switching state can be replaced by the following four states. (1) All of B1, B2, and B4 can open appropriately. (2) Only B1 is stuck. (3) Only B4 is stuck. (4) Only B6 is stuck. Apparently, the conditions with any two or all of the three breakers stuck simultaneously have been ignored since these events have an extremely low probability of occurrence. This case is shown in Figure 5.9. The transition rate from the up state to each of the four states is the active failure rate weighed by the probability of each stuck breaker event. For example, P_1 is the probability that B1 is stuck

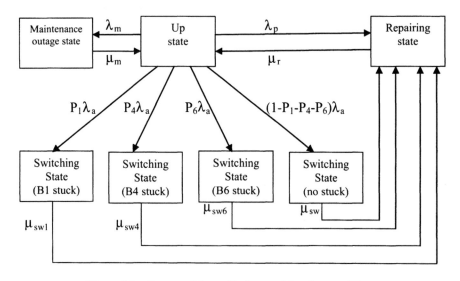

Figure 5.9 State model considering stuck breaker conditions.

but B4 and B6 are not stuck. P_4 or P_6 have similar definitions. Each of the stuck breaker states is associated with the outage of more healthy components. In this example, the B1 stuck state requires the opening of the breaker at the other end of the line and this generally will not create a worse impact on the reliability of the substation under assessment. The B4 or B6 stuck state requires the opening of breakers B2, B5, and B7, leading to the loss of the loads at buses 7 and 8. In general, subsequent switching actions may restore the power supply to some or all of the load points. Incorporation of stuck breaker conditions results in complexity in modeling. This is mainly due to the fact that an active failure of different components is associated with switching actions of different breakers and the number of breakers in the switching actions is varied. It is relatively easy to handle the complexity in the Monte Carlo methods but some simplifications are needed if an analytical technique is used. Figure 5.9 is just an example of considering the stuck breaker conditions for the short-circuit fault on T1 in the substation configuration shown in Figure 5.7. Similar state space diagrams can be drawn for other stuck breaker conditions.

5.4.2 Connectivity Identification

Once a system state of a substation configuration is selected using the state enumeration or Monte Carlo simulation method, a connectivity analysis is performed to determine the load points that are disconnected from the power source points. This can be done using the direct labeling technique. Let us use the substation configuration in Figure 5.7 to illustrate the technique.

All the nodes in the substation configuration are numbered. A bus section is a node and the other equipment has two nodes at its ends. In this example, nodes 4 to 9 are the bus sections. Nodes 1, 2, and 3 are the source points and nodes 7, 8, and 9 are the load points. All the switches are normally closed and their failures are not considered in the following discussion for simplicity. It is not difficult to include the failures of the switches in an actual application since the concept discussed is general to any failure. All the branches are put in a set, as shown in Figure 5.10 (a). For each system state, the connectivity between each source node and all the load nodes is examined. For instance, consider the system state caused by the short-circuit fault on T1 as an example. This corresponds to the switching state in Fig. 5.8. The procedure includes the following steps:

1. The branches that are out of service are crossed out from the set to form a residual set. In the fault state considered, the branches (1–4), (4–5), (4–7), and (7–8) are crossed off, as shown in Figure 5.10 (b).
2. Each source node is checked. If a source node does not remain in the residual set, this indicates that no connection exists between the source node and all the load nodes. This is the case for the source node 1 in the fault state considered.
3. If a source node remains in the residual set, the source node is first labeled. All the nodes that are directly connected to the labeled nodes can be further

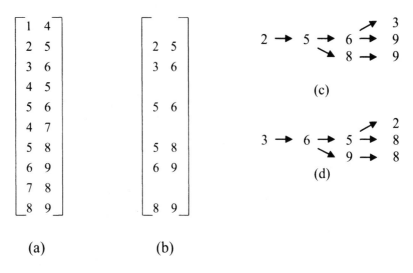

Figure 5.10 Connectivity identification for the system state caused by a short circuit on T1 in Figure 5.7.

labeled. Starting from the first labeled node, the labeling process is continued until it can no longer proceed. All the labeled nodes form a node set. If a load node is not found in the labeled node set, it does not have a connection to the source node. Otherwise, there is a connection between them. For the fault state considered, Figure 5.10 (c) and (d) show the labeling processes starting from the source nodes 2 and 3, respectively, and the corresponding labeled node sets. It can be asserted from Figure 5.10 (c) and (d) that there is no connection between the source node 2 (or 3) and the load node 7, but a connection exists between the source node 2 (or 3) and the load node 8 or 9.

4. If a load node has been disconnected from all the source nodes, this load node has a failure (load curtailment). This is the case for the load node 7 in the state examined.

After B1, B4, and B6 are opened by the protective action following the short circuit on T1, the two switches of T1 (indicated by S in Figure 5.7) can be opened manually or by using a supervisory control to isolate T1. Then B1, B4, and B6 can be reclosed. This leads to the repairing state in Figure 5.8, which will continue until T1 is repaired. For the connectivity identification in this state, the branches (1–4), (4–5), and (7–8) are added in the residual branch set. In examination of the connectivity between each source node and the load nodes, a similar labeling process creates the labeled node set that includes the load node 7. The connectivity identification process is illustrated in Figure 5.11 and shows that the repairing state does not require any load curtailment at any load point 7, 8, or 9.

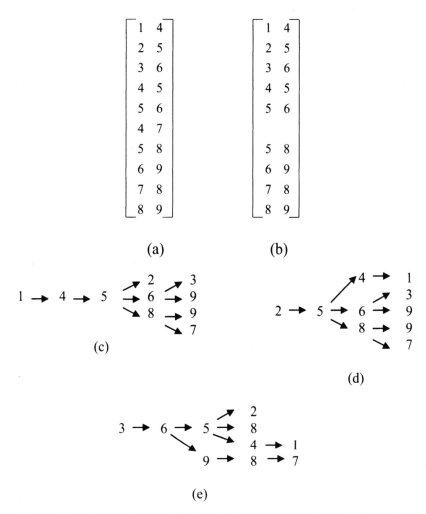

Figure 5.11 Connectivity identification for the system state after switching actions.

5.4.3 Stratified State Enumeration Method

The state enumeration method is often used in the risk evaluation of substation configurations. The number of system states increases exponentially with components and the number of component states. The following two measures are employed to reduce the burden in calculations:

1. The overlapping failure events are considered only to the second order. In other words, simultaneous outages of three or more components are neglect-

ed. This ignorance is rational since the probabilities of high-order failure events for substation components are extremely low.

2. The stuck breaker conditions are not represented in the state space diagram; a three-state (maintenance outage excluded) or a four-state (maintenance outage included) model is used. Impacts of the stuck breaker conditions are incorporated through an enumeration process of substates under the switching state. This approach basically does not create an error in the probability-related indices but an approximation in the frequency-duration indices.

The main steps of the stratified enumeration method are as follows:

1. A typical three-state or four-state model is applied to all substation components depending on whether the maintenance outage is considered or not. The switching state for each component is individually examined to determine:

   • Which healthy components are out-of-service due to an active failure
   • What switching actions are performed to isolate the failed component in order to transit to the repairing state
   • Which breakers/switches have a possible stuck condition with a probability of occurrence

2. The state space equations of all components are solved to obtain the probabilities of each state of the components.

3. The state enumeration technique is used to select system states. As mentioned earlier, only the first- and second-order failure events are considered. The system state probabilities are given by

$$P(si) = P_{si} \prod_{\substack{k=1 \\ k \neq i}}^{n} P_{uk} \qquad (i = 1, \ldots, n) \qquad (5.22)$$

$$P(ri) = P_{ri} \prod_{\substack{k=1 \\ k \neq i}}^{n} P_{uk} \qquad (i = 1, \ldots, n) \qquad (5.23)$$

$$P(si, sj) = P_{si}P_{sj} \prod_{\substack{k=1 \\ k \neq i,j}}^{n} P_{uk} \qquad (i, j = 1, \ldots, n; j \neq i) \qquad (5.24)$$

$$P(ri, rj) = P_{ri}P_{rj} \prod_{\substack{k=1 \\ k \neq i,j}}^{n} P_{uk} \qquad (i, j = 1, \ldots, n; j \neq i) \qquad (5.25)$$

$$P(si, rj) = P_{si}P_{rj} \prod_{\substack{k=1 \\ k \neq i,j}}^{n} P_{uk} \qquad (i, j = 1, \ldots, n; j \neq i) \qquad (5.26)$$

$$P(mi, sj) = P_{mi}P_{sj} \prod_{\substack{k=1 \\ k \neq i,j}}^{n} P_{uk} \qquad (i, j = 1, \ldots, n; j \neq i) \qquad (5.27)$$

$$P(mi, rj) = P_{mi} P_{rj} \prod_{\substack{k=1 \\ k \neq i,j}}^{n} P_{uk} \qquad (i, j = 1, \ldots, n; j \neq i) \tag{5.28}$$

Here, P_{uk} is the probability of the kth component in the up state, P_{si}, P_{ri}, and P_{mi} are the probabilities of the ith component in the switching, repairing, and maintenance states, respectively. $P(si)$ and $P(ri)$ are the probabilities of one component i being in the switching or repairing state, respectively, while all other components are in the up state. $P(si, sj)$ or $P(ri, rj)$ is the probability of two components i and j being in the switching or repairing state while all others in the up state. $P(si, rj)$ is the probability of one component i being in the switching state, another one j in the repairing state, and others in the up state. $P(mi, sj)$ or $P(mi, rj)$ is the probability of one component i being in the maintenance outage state, another one j in the switching or repairing state, and others in the up state. n is the number of the components in the substation configuration.

4. If the stuck breaker conditions are considered, a subenumeration process is performed. A system state associated with breaker switching is divided into several substates. Each substate corresponds to the one breaker stuck or no breaker stuck condition. The probabilities of the substates are calculated by

$$P_j(sb) = \begin{cases} P(st)\left[P_{bj} \displaystyle\prod_{\substack{k=1 \\ k \neq j}}^{m} (1 - P_{bk}) \right] & (j = 1, \ldots, m; \text{ Breaker } j \text{ stuck}) \\[3ex] P(st)\left\{ 1 - \displaystyle\sum_{i=1}^{m}\left[P_{bi} \displaystyle\prod_{\substack{k=1 \\ k \neq i}}^{m} (1 - P_{bk}) \right] \right\} & (j = 0; \text{ No breaker stuck}) \end{cases} \tag{5.29}$$

where $P(st)$ is the probability of the system being in the state that is associated with m switching breakers. It is the probability obtained using Equation (5.22), or (5.24), or (5.26), or (5.27) in Step 3. $P_j(sb)$ is the probability of the substate corresponding to the jth breaker being stuck or no breaker stuck ($j = 0$). P_{bj} or P_{bk} or P_{bi} is the probability of breaker j or breaker k or breaker i being stuck. It should be noted that the substates of two or more breakers being stuck simultaneously have been assumed to not occur and thus their probabilities have been included in the probability of no breaker being stuck. For this reason, the probability of no breaker being stuck in Equation (5.29) is expressed as 1.0 minus the sum of the probabilities of all the individual stuck breaker substates.

5. The connectivity between source points and load points is examined for each system state or its substate using the method described in Section 5.4.2. If a load point is disconnected from all the sources, this system state or substate is identified as a failure state for the load point and the curtailed load is recorded.

6. The risk indices at each load point are evaluated using the following equations.

   (a) Probability of load curtailments (PLC):

$$PLC_k = \sum_{i=1}^{N_k} P_{ik} \tag{5.30}$$

where P_{ik} is the probability of the ith failure state or substate associated with the load point k and N_k is the number of the failure states or substates in which the load at the load point k has to be curtailed.

(b) Expected demand not supplied (EDNS, MW):

$$EDNS_k = \sum P_{ik} L_k \tag{5.31}$$

where L_k is the average load (MW) at the load point k during the period considered, which is often one year in most cases.

(c) Expected frequency of load curtailments (EFLC, failures/yr). For each failure state of a load point, the state frequency can be obtained using the relationship between the state probability and frequency:

$$f_{ik} = P_{ik} \sum_{j=1}^{M} \lambda_j \tag{5.32}$$

where f_{ik} is the frequency of the ith failure state associated with the load point k. λ_j is the departure rate of the jth component in state i and can be a failure, repair, switching, maintenance, or recovery rate depending on the status of the component in state i. M is the number of the departure rates from state i.

The EFLC index is the cumulative failure frequency, which can be approximated by

$$EFLC_k = \sum_{i=1}^{N_k} f_{ik} \tag{5.33}$$

The transition frequencies between failure states should have been deducted from the sum of all failure state frequencies to obtain the accurate $EFLC_k$ index. However, the transition frequencies cannot be directly calculated using the state enumeration method. Ignorance of this deduction in the evaluation is acceptable from an engineering viewpoint since a transition between failure states only rarely takes place in a real operation process of substations.

(d) Average duration of load curtailments (ADLC, h/failure):

$$ADLC_k = \frac{PLC_k \cdot 8760}{EFLC_k} \tag{5.34}$$

It should be recognized that the above enumeration process applies to the average load at a load point for a given time period. If the load curve is modeled using the multiple step model as shown in Table 5.4, the total index is the sum of the products of the index for each load level and its probability.

5.4.4 State Duration Sampling Method

As in the distribution system risk evaluation, the main purpose of using the state duration sampling method is to simulate the probability distributions of component state duration and calculate the probability distributions of risk indices. The procedure includes the following steps:

1. The artificial operating cycle of each component is created through sampling the duration residing in each state shown in Figure 5.8. If the maintenance activity is a preplanned schedule, it can be treated as a deterministic event in the Monte Carlo simulation. In this case, the maintenance outage is temporarily excluded and considered in Step 2. For simplicity, the duration residing in each state is assumed to follow an exponential distribution in the following analysis. There is no difficulty in considering other distribution assumptions since the difference is only associated with how to generate a random variable from a particular distribution. The sample values of time to active failure (TTF_a), time to passive failure (TTF_p), time to switching (TTS), and time to repair (TTR) can be obtained using the inverse transform method of generating a random variable (see Appendices B.3 and B.4):

$$TTF_a = -\frac{1}{\lambda_a} \ln R_a \tag{5.35}$$

$$TTF_p = -\frac{1}{\lambda_p} \ln R_p \tag{5.36}$$

$$TTS = -\frac{1}{\mu_{sw}} \ln R_s \tag{5.37}$$

$$TTR = -\frac{1}{\mu_r} \ln R_r \tag{5.38}$$

where λ_a, λ_p, μ_{sw}, and μ_r are the active failure rate, passive failure rate, switching rate, and repair rate of a component. R_a, R_p R_s, and R_r are the four independent and uniformly distributed random numbers between [0, 1]. Note that the sample value of the time residing in the up state is determined by

$$TTF = \min\{TTF_a, TTF_p\} \tag{5.39}$$

A sequence of sample values of state duration forms an operating history.
2. The component operating history is modified by imposing the down time due to a scheduled maintenance on it.
3. The operating history of the substation configuration is obtained by combining the individual modified component operating histories. Each system state

on the operating history of the substation configuration is examined. This includes the following:

- If there is any active failure, switching activities must be examined to determine which healthy components are removed from service with the active failure component.

- The connection identification procedure described in Section 5.4.2 is performed to determine which load points are disconnected from power source points. The duration of the system state with one or more disconnected load points contributes to the outage time of the disconnected load points.

4. The risk indices at each load point are calculated using the following equations (see Section 5.4.3 for the definitions of the indices):

$$PLC_k = \frac{\sum_{i=1}^{N_k} T_{dik}}{T_s} \tag{5.40}$$

$$EDNS_k = \frac{\sum_{i=1}^{N_k} T_{dik} L_{ik}}{T_s} \tag{5.41}$$

$$EFLC_k = \frac{N_k}{T_s} \tag{5.42}$$

$$ADLC_k = \frac{8760 \cdot \sum_{i=1}^{N_k} T_{dik}}{N_k} \tag{5.43}$$

where N_k is the number of the outages associated with load curtailments at the load point k in the simulation time span T_s (in years). Note that the N_k here should not be confused with the N_k in Equations (5.30), (5.31), and (5.33) since they are conceptually different. T_{dik} (in years) is the duration of the ith outage associated with the load point k and L_{ik} is the average load at the load point k during the ith outage.

The probability distribution of any risk index can be calculated since the outages in each sampling year for each load point can be identified from the operating history of the substation configuration. For example, the probability distribution of the EFLC index is given by

$$P_f(i) = \frac{n(i)}{N} \qquad (i = 0, 1, 2, \ldots) \tag{5.44}$$

where N is the number of years in the whole simulation span, $n(i)$ indicates the number of the years in each of which there are i outages, and $P_f(i)$ is the probability that i outages take place within one year. The probability distributions of other indices can be obtained using the similar idea.

The procedure discussed above is based on the component model shown in Figure 5.8. The state duration sampling method can be easily extended to the more complex model shown in Figure 5.9. The difference is that an additional sampling process to simulate the stuck breaker conditions has to be added. When a component operating history is created, the relevant information associated with the switching actions is recorded. A uniformly distributed random number sequence between [0, 1] is generated for each breaker that has to act in the switching. If the random number is larger than the probability of the stuck condition, the breaker functions successfully for this particular sample. Otherwise, it is stuck. Whenever a stuck condition occurs, more switching actions are introduced accordingly, based on the protection coordination. This process enables one to modify the status of components (in-service or out-of-service) in the operating history of the substation configuration. The key to the method is to assure a clear understanding of each state of the substation configuration—which component really fails, which component is forced out of service just due to the active failure of another component, and which component is returned to service after switching actions.

5.5 TECHNIQUES USED IN COMPOSITE GENERATION AND TRANSMISSION SYSTEMS

Figure 5.12 shows a simple composite system. The generations and loads in the system are located at the different points and linked through the transmission network. In such a system, loads may still be curtailed even if all the generators are available. This is not only because one or more isolated buses may be created in a case of multiple line failures but also due to the fact that a failure event may cause overloading or voltage violations and force load curtailments. In other words, the system analy-

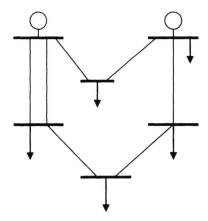

Figure 5.12   A simple composite system.

sis for composite system risk evaluation is not a simple connectivity problem. It is associated with the power flow, contingency analysis, and remedial actions such as overload alleviation, generation rescheduling, load curtailment philosophy, and switching measures. Many considerations in system state selection lead to additional complications in the evaluation. These include independent outages of system components, common cause, station-related and other dependent outages, weather effects, bus load uncertainty and correlation, derated state modeling, and other constraints in the system [31, 42–49].

Depending on different requirements and purposes, there are three possible cases in composite system risk evaluation:

1. The failures of both generation and transmission components are considered. This is a general case.
2. Only the failures of transmission components are considered, whereas generating units are assumed to be 100% reliable. This is the transmission system risk evaluation. In this case, some of substation arrangements may be incorporated in the evaluation for a special study purpose.
3. Only the failures of generating units are considered, whereas transmission components are assumed to be 100% reliable. It should be appreciated that this case is not the same as the generation-demand system risk evaluation since the constraints due to the transmission network are still included in the assessment.

5.5.1 Basic Procedure

Composite system risk evaluation has four main aspects: determination of component failures and load curve models, selection of system states, identification and analysis of system problems, and calculations of reliability indices. Both the state enumeration and Monte Carlo simulation methods can be applied to composite system risk evaluation. The two methods use different approaches to select system states and have different forms of formulas to calculate risk indices. The techniques of identifying and analyzing problems in a system state are the same. These include the power flow and contingency analysis for problem recognition and the optimal power flow for remedial actions. The basic procedure is shown in Figure 5.13.

5.5.2 Component Failure Models

Generating units are represented using the two-state (up and down) or multiple-state (derated states included) model. Generally, there are several generators at one bus. When the Monte Carlo method is used, the states or state transitions of all generators can be directly sampled and no simplification is needed. When the state enumeration method is utilized, the number of system states increases exponentially with the number of generators and the number of their derated states. To reduce the computational effort, a generation capacity probability table for each generator bus, which is similar to Table 5.1, can be created. Instead of using the state probabilities of individual generators, the probabilities of generation capacity levels in the table are enumerated.

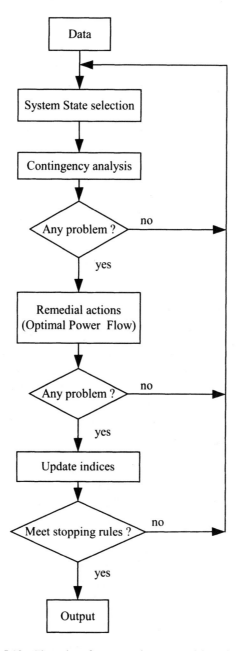

Figure 5.13   Flow chart for composite system risk evaluation.

Transmission components include overhead lines, cables, transformers, capacitors, and reactors. In general, these components are represented using the two-state (up and down) model. A multiple-state model including one or more derated states sometimes may be needed for a HVDC line. The common-cause outage model is used for overhead lines on the same right of way or a set of components controlled by the same protection logic. The tap-connection structure in a transmission line should be represented using the group outage model. Beside, the environment-dependent outages, cascading failures and station-originated outages can be modeled depending on special requirements.

The planned outages of some or all components can be considered when necessary. If the age of components in a system is a concern, their aging-failure model should be incorporated in the evaluation.

The details of the component models have been discussed in Chapter 2.

5.5.3   Load Curve Models

When the sequential Monte Carlo method is used, a chronological load curve is directly employed as the load model. For the state enumeration or state sampling method, a nonchronological load duration curve is utilized. There are three possible situations for the load model:

1. A single load curve is considered and loads at all buses are scaled proportionally to follow the shape of the given load curve. In this case, a multiple-step model that is similar to Figure 5.2 can be created to represent the load duration curve.
2. Loads at some buses may stay flat all the time, such as an industrial customer with the same power demand every hour. It is easy to model the constant loads at some buses, whereas other bus loads follow a load curve.
3. Loads are classed into the different bus groups that follow varied load curves. In this case, several load duration curves have to be considered and each curve represents a bus group. The multiple step load model has to capture the correlation between all the load curves.

A cluster technique is used to create the multiple-step load model for the third situation. Assume that the load duration curves are divided into NL load levels in a percentage of the peak. This corresponds to grouping the 8760 loads into NL clusters. Each load level is the mean value of the load points in a cluster for each curve. The technique includes the following steps:

Step 1: Select the initial values of the cluster means M_{ij}, where i and j denote cluster i ($i = 1, \ldots, NL$) and curve j ($j = 1, \ldots, NC$).

Step 2: Calculate the Euclidean distances from each hourly load point to each cluster mean by

$$D_{ki} = \left[\sum_{j=1}^{NC} (M_{ij} - L_{kj})^2 \right]^{1/2} \qquad (5.45)$$

where D_{ki} is the Euclidean distance from the kth load point to the ith cluster mean, L_{kj} the kth load value in the jth curve, and NC the number of curves.

Step 3: Regroup the load points by assigning them to the nearest cluster and calculate the new cluster mean by

$$M_{ij} = \sum_{k=1}^{N_i} L_{kj}/N_i \qquad (5.46)$$

where N_i is the number of the load points in the ith cluster.

Step 4: Repeat Steps 2 and 3 until all the cluster means remain unchanged between iterations.

The converged cluster means M_{ij} are used as the load levels for each curve in the multiple-step load model. In the actual programming, calculating the Euclidean distances and regrouping the load points are necessary only for two neighboring clusters if all the load points are arranged in descending order at the beginning.

5.5.4 Contingency Analysis

The analysis for generating unit contingencies is straightforward. If the remaining generation capacity at each generator bus can compensate the unavailable capacity due to loss of generator(s) at the same bus, there is no need for load curtailment. Otherwise, a generation rescheduling should be performed using the optimal power flow model that is discussed in the next section.

The transmission contingency analysis is more complex, The purpose is to calculate line flows and bus voltages following one or more component failures and identify if there is any overloading, voltage violation, isolated bus, or split island. Described below are the two basic transmission contingency analysis techniques that are often used in composite system risk evaluation.

5.5.4.1 AC Power-Flow-Based Sensitivity Technique. Consider a case of the line i–j outage in a transmission system. Assume that the flows on the two ends of the line i–j before its outage are $P_{ij} + jQ_{ij}$ and $P_{ji} + jQ_{ji}$ and image that there are two additional power injections at the buses i and j in the preoutage state, which are denoted by $\Delta P_i + j\Delta Q_i$ and $\Delta P_j + j\Delta Q_j$. If the additional power injections can produce power flow increments so that the power flows on the system are the same as those in the postoutage state, the effect of the additional power injections is completely equivalent to the outage of the line i–j. It can be proved that the power flows on the line i–j and the additional power injections in the preoutage state have the following relationship [1]:

$$
\begin{bmatrix} P_{ij} \\ Q_{ij} \\ P_{ji} \\ Q_{ji} \end{bmatrix} = \left[\begin{bmatrix} 1 & 0 & 0 & 0 \\ 0 & 1 & 0 & 0 \\ 0 & 0 & 1 & 0 \\ 0 & 0 & 0 & 1 \end{bmatrix} - \begin{bmatrix} \dfrac{\partial P_{ij}}{\partial P_i} & \dfrac{\partial P_{ij}}{\partial Q_i} & \dfrac{\partial P_{ij}}{\partial P_j} & \dfrac{\partial P_{ij}}{\partial Q_j} \\[2mm] \dfrac{\partial Q_{ij}}{\partial P_i} & \dfrac{\partial Q_{ij}}{\partial Q_i} & \dfrac{\partial Q_{ij}}{\partial P_j} & \dfrac{\partial Q_{ij}}{\partial Q_j} \\[4mm] \dfrac{\partial P_{ji}}{\partial P_i} & \dfrac{\partial P_{ji}}{\partial Q_i} & \dfrac{\partial P_{ji}}{\partial P_j} & \dfrac{\partial P_{ji}}{\partial Q_j} \\[2mm] \dfrac{\partial Q_{ji}}{\partial P_i} & \dfrac{\partial Q_{ji}}{\partial Q_i} & \dfrac{\partial Q_{ji}}{\partial P_j} & \dfrac{\partial Q_{ji}}{\partial Q_j} \end{bmatrix} \right] \begin{bmatrix} \Delta P_i \\ \Delta Q_i \\ \Delta P_j \\ \Delta Q_j \end{bmatrix} \quad (5.47)
$$

The additional power injections at the buses i and j can be found by solving Equation (5.47). Then the increments of bus voltage magnitudes and angles due to the line i–j outage can be obtained by solving the following equation:

$$
[J] \begin{bmatrix} \Delta \delta \\ \Delta V/V \end{bmatrix} = [\Delta I] \quad (5.48)
$$

where $[J]$ is the Jacobian matrix of the power flow equations in the preoutage state. $[\Delta V/V]$ is the voltage magnitude increment subvector whose elements are $\Delta V_i/V_i$, $[\Delta \delta]$ is the voltage angle increment subvector whose elements are δ_i, and $[\Delta I]$ is defined as follows:

$$
[\Delta I] = [0, \ldots, 0, \Delta P_i, 0, \ldots, 0, \Delta P_j, 0, \ldots, 0, \Delta Q_i, 0, \ldots, 0, \Delta Q_j, 0, \ldots, 0]^T
$$

Once the bus voltages are obtained, the line power flows following the line i–j outage can be calculated. A similar procedure can be applied to multiple line outages.

5.5.4.2 DC Power-Flow-Based Contingency Analysis.
DC power-flow-based contingency analysis provides fast and sufficiently accurate real power flows following line outages for the risk evaluation in which a huge number of outage events have to be considered.

The bus impedance matrices after multiple line outages can be calculated directly from the bus impedance matrix before the outages:

$$
Z(S) = Z(0) + Z(0)MQM^T Z(0) \quad (5.49)
$$

$$
Q = [W - M^T Z(0)M]^{-1} \quad (5.50)
$$

where, $Z(0)$ and $Z(S)$ are the bus impedance matrices before and after the line outages, in which the resistance of each line is neglected. 0 and S in brackets denote the normal and outage system states, respectively. W is the diagonal matrix whose each diagonal element is the reactance of the outage line. M is the submatrix composed of the columns corresponding to the outage lines of the bus-line incidence matrix.

The line flows after the line outages can be calculated by

$$T(S) = A(S)(PG - PD) \tag{5.51}$$

where, $T(S)$ is the real power flow vector in the outage state. PG and PD are the generation output and load power vectors, respectively. $A(S)$ is the relation matrix between the real power flows and power injections in the outage state S. The mth row of $A(S)$ can be calculated as follows:

$$A_m(S) = \frac{Z_r(S) - Z_q(S)}{X_m} \tag{5.52}$$

where, X_m is the reactance of the mth line, the subscripts r and q denote the two bus numbers of the mth line, and $Z_r(S)$ and $Z_q(S)$ are the rth and qth rows of $Z(S)$, respectively.

5.5.5 Optimization Models for Load Curtailments

When an outage causes system problems, a special optimal power flow (OPF) model is used to reschedule generations and alleviate constraint violations, and at the same time to avoid any load curtailment if possible or to minimize the total load curtailment if unavoidable. The objective function of the OPF model is minimization of the total load curtailment, whereas load curtailments at buses are the solutions of the model. The risk indices are based on the load curtailments in selected system outage states and their probabilities of occurrence.

Corresponding to the two contingency analysis techniques, the two optimal power flow models are discussed below.

5.5.5.1 AC Power-Flow-Based OPF Model. The AC power-flow-based OPF model for composite system risk evaluation can be described as follows:

$$\min \sum_{i \in ND} C_i \tag{5.53}$$

subject to

$$P_i(V, \delta) - PD_i + C_i = 0 \quad (i \in ND) \tag{5.54}$$

$$Q_i(V, \delta) - QD_i = 0 \quad (i \in ND) \tag{5.55}$$

$$PG_i^{\min} \le P_i(V, \delta) \le PG_i^{\max} \quad (i \in NG) \tag{5.56}$$

$$QG_i^{\min} \le Q_i(V, \delta) \le QG_i^{\max} \quad (i \in NG) \tag{5.57}$$

$$0 \le C_i \le PD_i \quad (i \in ND) \tag{5.58}$$

$$T_k(V, \delta) \le T_k^{\max} \quad (k \in L) \tag{5.59}$$

$$V_i^{\min} \le V_i \le V_i^{\max} \quad (i \in N) \tag{5.60}$$

where, P_i and Q_i are the real and reactive power injections at bus i; V and δ are the bus voltage magnitude and angle vectors; V_i is an element of V; PD_i and QD_i are the real and reactive loads at bus i; C_i is the load curtailment variable at bus i; $PG_i^{\min}$, $PG_i^{\max}$, $QG_i^{\min}$, and $QG_i^{\max}$ are the lower and upper limits, respectively, of the real and reactive power injections at the generation bus i; T_k is the MVA power flow on line k; $T_k^{\max}$ is the rating limit of line k; $V_i^{\min}$ and $V_i^{\max}$ are the lower and upper limits of the voltage magnitude at bus i; ND, NG, N and L are, respectively, the sets of load buses, generator buses, all buses, and all circuits in the system. The P_i, Q_i, and T_k have the following expressions, respectively:

$$P_i(V, \delta) = V_i \sum_{j \to i} V_j(G_{ij} \cos \delta_{ij} + B_{ij} \sin \delta_{ij}) \tag{5.61}$$

$$Q_i(V, \delta) = V_i \sum_{j \to i} V_j(G_{ij} \sin \delta_{ij} - B_{ij} \cos \delta_{ij}) \tag{5.62}$$

where, $j \to i$ denotes that j are the buses directly connected to bus i through lines. G_{ij} and B_{ij} are the real and imaginary parts of the ith row and jth column element of the bus admittance matrix; δ_{ij} is the angle difference, that is, $\delta_{ij} = \delta_i - \delta_j$.

$$T_k(V, \delta) = \max\{T_{mn}(V, \delta), T_{nm}(V, \delta)\} \tag{5.63}$$

where, $T_{mn}(V, \delta)$ and $T_{nm}(V, \delta)$ are the MVA flows at the two ends of Line k. The m and n are the two buses of line k. The MVA flow from bus m to n is calculated by

$$T_{mn}(V, \delta) = \sqrt{P_{mn}^2(V, \delta) + Q_{mn}^2(V, \delta)} \tag{5.64}$$

$$P_{mn}(V, \delta) = V_m^2(g_{mo} + g_{mn}) - V_m V_n(b_{mn} \sin \delta_{mn} + g_{mn} \cos \delta_{mn}) \tag{5.65}$$

$$Q_{mn}(V, \delta) = -V_m^2(b_{mo} + b_{mn}) + V_m V_n(b_{mn} \cos \delta_{mn} - g_{mn} \sin \delta_{mn}) \tag{5.66}$$

where, $g_{mn} + jb_{mn}$ is the circuit admittance of line k and $g_{mo} + jb_{mo}$ is the equivalent admittance of the circuit to the ground at the end of bus m.

It can be seen that the OPF model minimizes the total load curtailment while satisfying the AC power flow equations and line flow, voltage, and generation output limits. It is a nonlinear programming problem and can be solved using the interior point method, which is described in Appendix D.2.

5.5.5.2 DC Power-Flow-Based OPF Model.
In the risk evaluation of a composite system, there may be a case in which thousands of system states require solving an optimal power flow model. To reduce the computational burden, the DC

power-flow-based OPF model is often used. It can be mathematically expressed as follows:

$$\min \sum_{i \in ND} C_i \tag{5.67}$$

subject to

$$T(S) = A(S)(PG - PD + C) \tag{5.68}$$

$$\sum_{i \in NG} PG_i + \sum_{i \in ND} C_i = \sum_{i \in ND} PD_i \tag{5.69}$$

$$PG_i^{\min} \le PG_i \le PG_i^{\max} \quad (i \in NG) \tag{5.70}$$

$$0 \le C_i \le PD_i \quad (i \in ND) \tag{5.71}$$

$$|T_k(S)| \le T_k^{\max} \quad (k \in L) \tag{5.72}$$

where, $T(S)$, $A(S)$, PG, and PD are the same as defined in Equation (5.51). C is the load curtailment vector. PG_i, PD_i, C_i, and $T_k(S)$ are the elements of PG, PD, C, and $T(S)$, respectively. $PG_i^{\min}$, $PG_i^{\max}$, and $T_k^{\max}$ are the limits of PG_i and $T_k(S)$, respectively. NG, ND, and L are the sets of generation buses, load buses, and circuits in the system. The objective of the model is still to minimize the total load curtailment while satisfying the power balance, DC power flow relationships, and limits on line flows and generation outputs. The algorithm for this linear programming model is provided in Appendix D.1.

Compared to the AC power-flow-based OPF model, all the quantities associated with reactive power have been ignored in the DC model. Considerable calculations indicate that this is reasonably acceptable for composite system risk evaluation.

5.5.6 State Enumeration Method

State enumeration is one of the techniques used in composite system risk evaluation, particularly in the cases where component failure probabilities are relatively low. The basic concept of the state enumeration method has been described in Section 4.3.1. The procedure for composite system risk assessment includes the following aspects.

1. Create the multiple-step load model as described in Section 5.5.3. The state enumeration is performed for each load level.
2. Select a system state using the enumeration technique. The probability of the system state is calculated by

$$P(s) = \prod_{i=1}^{n_d} PF_i \prod_{i=1}^{n_r} PP_i \prod_{i=1}^{n-n_d-n_r} (1 - PF_i - PP_i) \tag{5.73}$$

where n_d and n_r are the numbers of components unavailable (down state) and partially unavailable (derated state) in the system state s, n the total number of components in the system, PF_i the unavailability of component i, and PP_i the probability of component i in the derated state.

3. Perform a contingency analysis and, if there is a system problem, conduct an optimal power flow analysis to evaluate the load curtailment required in the selected state. If the load curtailment is not zero, the selected state is a failed one. The probabilities and load curtailments of only failure system states are recorded.

4. Calculate the risk indices using the following equations:

   (a) Probability of load curtailments (PLC):

$$PLC = \sum_{i=1}^{NL} \left(\sum_{s \in F_i} P(s) \right) \frac{T_i}{T} \tag{5.74}$$

where, $P(s)$ is the probability of state s, F_i the set of all the failure system states at the ith load level in the multiple-step load model, T_i the time length (in hours) of the ith load level, and NL the number of load levels. T is the total time period of the load curve (in hours). It is often one year for system development planning and can be any time length (week, month, or season) for operation planning.

   (b) Expected energy not supplied (EENS, MWh/period):

$$EENS = \sum_{i=1}^{NL} \left(\sum_{s \in F_i} P(s) \cdot C(s) \right) \cdot T_i \tag{5.75}$$

where, $C(s)$ is the load curtailment (MW) in state s.

   (c) Expected frequency of load curtailments (EFLC, failures/period):

$$EFLC = \sum_{i=1}^{NL} \sum_{s \in F_i} \left(P(s) \sum_{j=1}^{m(s)} \lambda_j \right) \frac{T_i}{T} \tag{5.76}$$

where, λ_j is the jth departure rate of the components in state s. $m(s)$ is the total number of the transition rates departing from state s. Note that the $m(s)$ may be different from the number of components. If there is one or more components that are represented by a multiple-state model (such as derated states of generators), $m(s)$ is larger than the number of components. It should be recognized that the transition frequencies between failure states have been ignored in the equation, which gives the approximate estimation of the frequency index.

   (d) Average duration of load curtailments (ADLC, h/failure):

$$ADLC = \frac{PLC*T}{EFLC} \tag{5.77}$$

All the four equations can be used to calculate the indices for individual buses or the overall system. For the bus indices, F_i is the set of the system states only associated with the load curtailments at a particular bus, whereas for the system indices, it is the set of the system states associated with the load curtailments at any bus. Note that the PLC or EFLC or ADLC index for the system is not the sum of the indices for all the buses. If a failure system state requires load curtailments at multiple buses, it can be counted only once in calculating the system PLC or EFLC or ADLC index while it makes contributions to the index for each of all the buses with the load curtailment. The system EENS index should be identical to the sum of the EENS indices of all the buses because the load curtailments at the buses in any failure system state sum to the total curtailment for the system.

5.5.7   State Sampling Method

The state sampling method is often used in composite system risk evaluation, particularly in cases where complex operational measures and/or some special considerations such as the bus load uncertainty and correlation or weather impacts need to be simulated [43, 44]. The concept of the state sampling for composite system risk assessment is basically the same as that for generation system risk evaluation, which has been described in Section 5.2.2. Sampled components should cover both transmission equipment and generating units. An essential difference is the evaluation process of load curtailments. The contingency analysis and optimal power flow must be conducted in composite system risk evaluation to identify system failure states and calculate load curtailments. The procedure includes the following aspects.

1. Create the multiple step load model as described in Section 5.5.3. The state sampling process is performed for each load level.
2. Select a system state using the Monte Carlo simulation. This is associated with random determination of bus loads and component states.

    (a) The normally distributed random variables are used to capture uncertainties of bus loads:

$$M_{\sigma ij} = (z_{ij}\sigma_{ij} + 1)M_{ij} \tag{5.78}$$

   where, M_{ij} is the ith load level on curve j in the multiple-step load model obtained using the clustering approach in Section 5.5.3. $M_{\sigma ij}$ is a sample value of the bus load following curve j at the ith load level. Both M_{ij} and $M_{\sigma ij}$ are expressed in the per-unit value with respect to the peak. z_{ij} is the standard normal distribution random number corresponding to the ith load level i on curve j and σ_{ij} is the standard deviation representing the bus load uncertainty in a percentage of M_{ij}.

The megawatt sample value of the bus load that follows the load curve j is obtained by using $M_{\sigma ij}$ times its peak megawatts. If the correlation between bus loads is considered, a correlation sampling approach is applied [44].

(b) Component states (up, down or derated) are simulated using the uniformly distributed random variables:

$$s_k = \begin{cases} 0 \text{ (up)} & \text{if } R_k > PP_k + PF_k \\ 1 \text{ (down)} & \text{if } PP_k < R_k \le PP_k + PF_k \\ 2 \text{ (derated)} & \text{if } 0 \le R_k \le PP_k \end{cases} \qquad (5.79)$$

where, s_k is the sampled state of component k; R_k is a uniformly distributed random number between $[0, 1]$ for the kth component; and PF_k and PP_k are the probabilities of the down and derated states for the kth component.

3. Perform a contingency analysis and, if necessary, conduct an optimal power flow analysis to evaluate the load curtailment required in the selected state. If the load curtailment is not zero, the selected state is a failed one. The load curtailments in failure system states are recorded. Obviously, this step is exactly the same as the one used in the enumeration method.

4. Calculate the risk indices using the following equations.

(a) Probability of load curtailments (PLC):

$$PLC = \sum_{i=1}^{NL} \left(\sum_{s \in F_i} \frac{n(s)}{N_i} \right) \frac{T_i}{T} \qquad (5.80)$$

where, $n(s)$ is the number of state s occurring in the sampling, N_i the total number of samples, and F_i the set of all the failure system states at the ith load level in the multiple-step load model. T_i is the time length (in hours) of the ith load level, T the total time period (in hours) of the load curve, and NL the number of load levels.

(b) Expected energy not supplied (EENS, MWh/period):

$$EENS = \sum_{i=1}^{NL} \left(\sum_{s \in F_i} \frac{n(s)C(s)}{N_i} \right) T_i \qquad (5.81)$$

where, $C(s)$ is the load curtailment (MW) in state s.

(c) Expected frequency of load curtailments (EFLC, failures/period):

$$EFLC = \sum_{i=1}^{NL} \sum_{s \in F_i} \left(\frac{n(s)}{N_i} \sum_{j=1}^{m(s)} \lambda_j \right) \frac{T_i}{T} \qquad (5.82)$$

where, λ_j is the jth departure rate of the components in state s. $m(s)$ is the total number of the transition rates departing from state s. Similar to Equation (5.76), the transition frequencies between failure states have been ignored in Equation (5.82), leading to the approximate estimation of the frequency index.

(d) Average duration of load curtailments (ADLC, hours/failure). The equation for ADLC is the same as Equation (5.77).

It can be seen that the difference between the equations of the indices using the enumeration and state sampling methods is only associated with how to evaluate the probabilities of system states. Similarly, all the above equations can be used to calculate the indices for individual buses and the system. The key is to distinguish the set F_i of the failure system states, which is different for each bus and the overall system.

5.6 CONCLUSIONS

This chapter presents the risk evaluation techniques for power systems. These techniques are the further developments of the general methods described in Chapter 4. The details of the system state selection and system analysis techniques for generation, distribution, substation, and composite generation and transmission systems have been discussed.

The risk assessment of generation systems is mathematically a convolution problem. Although the probability convolution method can be directly used, the Monte Carlo simulation techniques provide an easier tool to deal with the load curves and load uncertainties. Essentially, a radial distribution system is a series network. The analytical technique based on the series network concept has been widely used in the risk evaluation of radial distribution systems. The state duration sampling method can simulate the probability distributions of input data and create the probability distributions of risk indices. This advantage makes it superior to the series network method. The labeling approach has been developed to perform the connectivity identification for an outage state in substation configurations. It can be used in both the stratified state enumeration and state duration sampling methods for substation system risk evaluation. One of the keys in the substation risk analysis is the modeling of the component failure modes and stuck breaker conditions. This is associated with the switching logic, which is generally varied in different substation layouts. The risk assessment of composite generation and transmission systems is the most complicated one, mainly due to the complexity in the system analysis. It requires the use of the power flow, contingency analysis, and optimal power flow models. Both the AC and DC power-flow-based contingency analyses and optimization techniques have been illustrated.

The risk indices are the numerical results of power system risk assessment. The calculation formulas of the indices for generation, distribution, substation, and composite generation and transmission systems have been developed. The formulas using the analytical and Monte Carlo simulation techniques have different expressions. There is no doubt, however, that they should provide consistent answers to the same problem.

It is important to recognize that only the bare bones of the techniques are addressed in this chapter. Many considerations in actual applications require extensions of the basic techniques and these will be presented in accordance with their different applications in the following chapters.

CHAPTER 6

APPLICATION OF RISK EVALUATION TO TRANSMISSION DEVELOPMENT PLANNING

6.1 INTRODUCTION

The fundamental task of transmission planning is to develop the system as economically as possible and maintain an acceptable reliability level. This is associated with determination of a reinforcement alternative and its implementation time. The deterministic $N - 1$ principle—the single-contingency criterion—has been widely used in the planning practice of utilities for many years. The principle requires that an outage of one single system component does not cause any damage or load curtailment. The $N - 1$ criterion has two weaknesses. First, multiple component failures are excluded from consideration. Second, only the outcomes of single-component failure events are analyzed but their probabilities of occurrence have been neglected. A failure event, even if extremely undesirable, is of little consequence if it is so unlikely that it can be ignored. Planning alternatives based on such an analysis will lead to overinvestment. Conversely, if selected failure events are not very severe but have relatively high probabilities of occurrence, an alternative based on the deterministic analysis of such events will still result in a high risk. Probabilistic risk assessment can capture multiple-component failure events and recognize not only the severity of the events but also the likelihood or probability of their occurrence. Appropriate combination of both severity and likelihood creates the indices that truly represent system risk.

The purpose of applying probabilistic risk analysis is to add one more dimension in system planning but not to replace the $N - 1$ principle, which has been accepted by system planners for years. There is no conflict between the $N - 1$ principle and risk evaluation. Figure 6.1 gives a conceptual example in which seven candidate

Risk Assessment of Power Systems. By Wenyuan Li
ISBN 0-471-63168-X © 2005 the Institute of Electrical and Electronics Engineers, Inc.

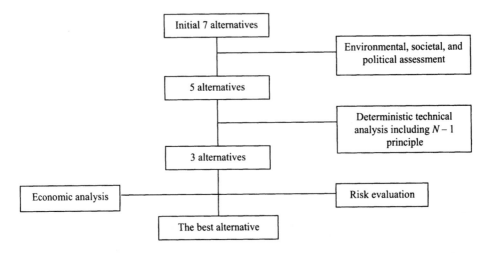

Figure 6.1   Outline of the system planning process.

planning alternatives are assumed at the beginning. Two of them are excluded based on environmental, societal, or political considerations. The deterministic technical criteria, including the $N-1$ principle, are applied to the remaining five alternatives. Two more alternatives are eliminated from the candidate list due to inability to meet the deterministic contingency criterion. Then a combined economic and risk analysis is performed to select the best scenario. Both the $N-1$ principle and the risk analysis can be applied. A system planning process includes societal, environmental, technical, and economical assessments, and risk evaluation can become a part of the whole assessment process.

The intent of this chapter is not to cover all the aspects of the planning process. The focus is on the application of risk evaluation in transmission development planning, which is often called probabilistic planning. The concept of probabilistic planning is discussed first in Section 6.2. The risk evaluation approach with the details of a risk cost model is described in Section 6.3. Two actual examples are given in Sections 6.3 and 6.4. The first one is associated with selection of the lowest total cost alternative and the second one provides a demonstration of applying two different planning criteria.

6.2   CONCEPT OF PROBABILISTIC PLANNING [32, 34, 35, 37]

6.2.1   Basic Procedure

Transmission probabilistic planning includes the following steps:

1. If the single-contingency criterion is a mandate, select the planning alternatives of meeting the $N-1$ principle using a traditional load flow or contin-

gency analysis program. If the $N-1$ principle is not considered a strict criterion, select all feasible alternatives.

2. Conduct risk evaluation and risk cost evaluation for the selected alternatives over a planning time frame (such as 10 years), using a risk assessment tool for composite or transmission systems.

3. Calculate the cash flows and present values of investment, operation, and risk costs for the selected alternatives in the planning time period.

4. Apply the performance-based cost efficiency criterion. The best alternative should achieve the minimum total cost:

$$\min T = I + O + R \tag{6.1}$$

where, I, O, and R are the investment, operation and risk costs respectively.

The term "cost efficiency" means the minimum total cost and the term "performance-based" means that the system unreliability cost is evaluated and included in the total cost as the risk component. If the selected alternatives in Step 1 above include those not meeting the single-contingency criterion, this is the "pure" performance-based cost efficiency criterion. Obviously, in this case, the single-contingency principle is not an imperative requirement. The impacts of single contingencies combined with their probabilities of occurrence are considered as a part of the overall system risk. Otherwise, if those not meeting the single-contingency criterion have been excluded, this is the combined criterion of single contingency and cost efficiency.

Step 2 is the key to probabilistic planning and will be discussed in detail in Section 6.3. The cost analysis approach for Step 3 is described first in the following subsection.

6.2.2 Cost Analysis

The cash flow of the annual investment cost is estimated using the capital return factor (CRF) [17, 18]:

$$A = V \cdot CRF \tag{6.2}$$

$$CRF = \frac{i(1+i)^n}{(1+i)^n - 1} \tag{6.3}$$

Here, A is the annual equivalent capital, V the actual investment in some year, i the discount rate, and n the economic life (years) of the investment V.

If the uncertainty of the discount rate is considered, a probability distribution of i can be introduced into Equation (6.3).

The cash flows of the operation and risk costs can be calculated through year-by-year evaluations. The operation cost in a transmission system is generally associated with network losses and can be obtained using energy price times energy losses; the latter is calculated using a power flow method considering a load duration curve. In many cases, all planning alternatives selected may be only involved in limited modifications in the network configuration and have basically the same network losses. In such a situation, the operation cost may not be necessarily included in the total cost for comparison.

The risk cost refers to the interruption damage cost due to random system component outages. Conceptually, it equals the product of the unit interruption cost ($/MWh) and the risk index EENS (expected energy not supplied, MWh/yr). There are different principles for estimation of the unit interruption cost. For instance, it can be the utility damage only, such as the revenue reduction or the additional cost due to emergency actions (e.g., running a diesel station in an outage situation); or it can be the interruption cost, including customer damages that can be obtained through a customer survey. The three general approaches to estimating the unit interruption cost have been discussed in Section 1.2.3. Which approach is used depends on individual cases and vary for different countries, utilities, and possibly even for different regions and projects in a utility. In applications, two methods can be used to obtain the risk cost. In the first one, the EENS is evaluated and multiplied by the unit interruption cost. In the second one, the unit interruption cost can be directly incorporated into the risk cost model associated with load curtailments.

6.2.3 Present Value

Two methods are used to calculate the total cost over a cash flow. The first one is to simply sum up the annual costs for all years on the cash flow in the planning time span. This approach does not differentiate the time values of the costs in different years. The second one is the present value method, which can capture the time values of the costs. In principle, both the methods are acceptable if the purpose is only the comparison between planning alternatives. The present value method is outlined here and used in the examples of this chapter.

The present value of the investment, operation, or risk cost is calculated using the following formula:

$$PV = \sum_{j=1}^{m} \frac{A_j}{(1 + i)^{j-1}} \qquad (6.4)$$

where, PV is the present value, A_j the annual cost in year j, i the discount rate, and m the number of years considered in system planning.

6.3 RISK EVALUATION APPROACH

Risk cost estimation is the key to total cost analysis. It requires a risk evaluation process.

6.3.1 Risk Evaluation Procedure

The Monte Carlo based risk evaluation approach is used in the two examples given in Sections 6.4 and 6.5. It includes the following aspects:

1. A multiple-step load model is created that eliminates the chronology and aggregates load states using hourly load records during one year. Annualized indices are calculated first by using only a single load level expressed on a one-year basis. All the load level steps are considered successively and the resulting indices for each load level are weighted by its probability to obtain annual indices.

2. The system states at a particular load level are selected using Monte Carlo simulation techniques. This includes the following:

   (a) Generally, generating-unit states are modeled using multiple-state random variables. If generating units do not create different impacts on selected transmission planning alternatives, the generating units can be assumed to be 100% reliable.

   (b) Transmission circuit states are modeled using the two-state (up and down) random variables. For some special transmission components such as HVDC lines, a multiple-state random variable can be applied. Weather-related transmission line forced outage frequencies and repair times can be determined using the method of recognizing regional weather effects (see Section 2.3.5). Transmission line common cause outages are simulated by separate random numbers.

   (c) The bus load uncertainty and correlation are modeled using a correlative normal distribution random vector. A tabulating technique of normal distribution sampling and a correlation sampling technique are used to select bus load states [1, 44].

3. The DC load flow-based system analysis approach is used to perform the contingency analysis and contingency state "filtering." The minimization model for load curtailments is solved only for those contingency states that may lead to load curtailments.

4. The linear programming minimization model is used to reschedule generations, alleviate line overloads, and avoid load curtailments if possible, or minimize the total damage cost if unavoidable. This model is described in the next section.

5. The risk indices are calculated. In probabilistic planning, the most important indices are the expected damage cost (EDC) and expected energy not supplied (EENS).

6.3.2 Risk Cost Model

The general optimization models for transmission risk assessment have been discussed in Section 5.5.5. In individual cases, modifications need to be introduced for different purposes.

For generality, two types of customers are considered: interruptible and firm loads. The interruptible loads, such as some industrial customers, can be shed in contingency situations in accordance with the agreement between the utility and customers. The damage cost due to this kind of load shedding is the revenue reduction for the utility. At the customer side, the damage cost is relatively small because the customers are prepared for power supply interruptions. The curtailments of firm loads are caused by forced outages without any prenotification. These curtailments are different by nature from the load shedding of the interruptible loads. The unit damage cost due to the firm load curtailments is much higher than that due to shedding the interruptible loads.

Loads at each bus are divided into interruptible and firm loads. For each contingency state that may lead to load shedding or curtailments, the following optimization model is used to reschedule generation outputs in order to alleviate line overloads while maintaining the generation–demand balance [34].

$$\min f = \sum_{i \in ND} W_i(d_s S_i + d_c C_i) \tag{6.5}$$

subject to:

$$T_k + \sum_{i=1}^{N} A_{ki}(PG_i - PD_i + S_i + C_i) \qquad (k \in L) \tag{6.6}$$

$$\sum_{i \in NG} PG_i + \sum_{i \in ND} (S_i + C_i - PD_i) = 0 \tag{6.7}$$

$$PG_i^{\min} \le PG_i \le PG_i^{\max} \qquad (i \in NG) \tag{6.8}$$

$$0 \le S_i \le R_i PD_i \qquad (i \in ND) \tag{6.9}$$

$$0 \le C_i \le (1 - R_i)PD_i \qquad (i \in ND) \tag{6.10}$$

$$-T_k^{\min} \le T_k \le T_k^{\max} \qquad (k \in L) \tag{6.11}$$

where, f is the weighted total damage cost ($/h), S_i and C_i are the two load curtailment variables (MW) corresponding to the interruptible and firm loads, respectively, at bus i; d_s and d_c are the unit damage costs ($/MWh) for interruptible and firm loads, respectively; W_i is the weighting factor reflecting bus load importance; PG_i and PD_i are the generation variable and load demand at bus i; T_k is the real power flow on line k; $PG_i^{\min}$, $PG_i^{\max}$, and $T_k^{\max}$ are the limit values of PG_i, and T_k, respectively; A_{ki} is an element of the relation matrix between real power flows and power injections; R_i is the interruptible load as a fraction of the total load at bus i; ND, NG, and L are the sets of load buses, generator buses, and lines, respectively; N is the number of buses in the system.

The objective of the linear programming model is to minimize the total damage cost of load shedding and curtailments satisfying the power balance, linearized load flow relationships, line power flow limits, and generation output limits. The model is solved using an algorithm combining the linear programming relaxation technique with the dual simplex method (see Appendix D.1).

The damage cost obtained using the above model corresponds to a particular system contingency state. The risk cost is the mean value of the damage costs for all system states sampled in the Monte Carlo simulation—the expected damage cost (EDC) index. This model also provides the EENS indices. Since two types of loads (interruptible and firm) are considered in the model, two EENS indices are created accordingly. One corresponds to the interruptible loads and another to the firm loads. The annual total risk cost can also be calculated by

$$EDC = d_s \cdot EENS(I) + d_c \cdot EENS\,(f) \qquad (6.12)$$

where, $EENS(I)$ and $EENS\,(f)$ are the annual expected energy not supplied for the interruptible and firm loads, respectively. For the planning alternative that is not associated with the interruptible load policy, the variable S_i can be taken away from the minimization model and the first term in Equation (6.12) is zero.

6.4 EXAMPLE 1: SELECTING THE LOWEST-COST PLANNING ALTERNATIVE [34, 51, 52]

Two actual applications using the regional systems of a utility are described in this and the next section to provide more details of the probabilistic planning concept.

6.4.1 System Description

Figure 6.2 shows the south metro 230 kV and 69 kV systems in a big city. The buses 619, 1201, and 1215 are three power injection points, and the bus 1201 is a major one. These are not actual generator buses but equivalent power injections. The metro system area is far away from major generation sources. In order to make the power flows to the south metro area realistic, the network constraint between the south metro area and generation sources has to be incorporated into the evaluation. The evaluated system includes the full south metro system, the 230 kV north metro system, and the 500 kV transmission network connecting the generation sources to the metro system. Some equivalent loads or power injections are used to represent other subsystems outside the metro area. No equivalence is used inside the south metro system. The planning task is to reinforce the south metro system. In order to highlight the risk in this area, component failures only in the south metro system are simulated and all the components outside this area are assumed to be 100% reliable. As a result, all load curtailments due to component outages are located inside the south metro system. The failure data is retrieved from the reliability database of the utility to which the studied system belongs. The study was performed in 1994. The planning time frame is the period of 7 years from 1997 to 2003. According to the load forecast, it is assumed that each year has 2% load growth. Also it is assumed

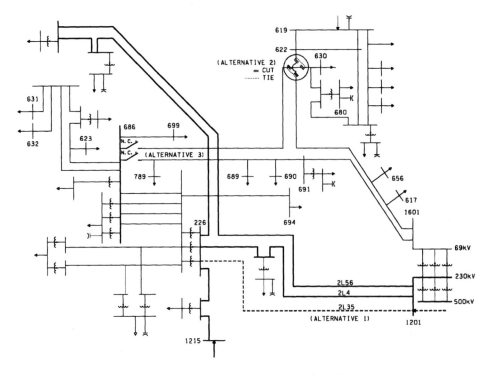

Figure 6.2 South metro system of a utility.

that the annual load duration curves for different years have the same shape. This means that 8760 hourly loads in a different year increase in the same proportion as the peak load in the year.

6.4.2 Planning Alternatives

The following four alternatives were considered in the study:

1. As load demands increase, overloading would be caused by some single outages in the south metro system. For the 1997 peak load level and beyond, when Line 2L4 (see Figure 6.2) trips out, the overload would appear on Line 2L56. Conversely, when Line 2L56 trips out, the overload would appear on 2L4. The initial reinforcement scheme is the addition of a 230 kV line (2L35) between Buses 226 and 1201 (Alternative 1). Such a consideration is natural since this line could share the power flow from Line 2L4 or Line 2L56. However, the alternative requires a considerable capital investment. Is there any other scheme that could provide the same reliability improvement but with a lower investment?

2. It is found by power flow analysis that a local configuration change ("cut and tie") in the 69 kV subsystem could avoid the overloading on 2L4 or 2L56

when either one trips out. The "cut-and-tie" scheme (Alternative 2) is shown in the figure. This alternative requires a much lower capital investment. On the other hand, however, the "cut and tie" creates higher loading levels on other lines in some system operating states even though it is beneficial for avoiding the overloading on 2L4 or 2L56.

3. Using a contingency analysis tool, it is found that when 2L4 (or 2L56) trips out and if the two 69 kV lines between Buses 686 and 630 and between Buses 686 and 789 are opened intentionally, the overloading on 2L56 (or 2L4) would disappear. This indicates that the south metro system has a feature of reliability noncoherence. The third alternative is operational manipulation: each time 2L4 or 2L56 fails and the overload occurs on the other line, the two 69 kV lines are intentionally opened (Alternative 3). They still remain normally closed in all other system operating states. This alternative requires revision of the operation guideline and installation of an automatic control and monitoring device. The cost of such a device, however, is very low and, therefore, it can be considered as a "noninvestment" option compared to the investment cost of Alternative 1.

4. There are eight industrial customers in the south metro area that are located at Buses 623, 630, 631 (632), 689, 690, 694, 699, and 789 (see Figure 6.2). In an emergency situation, the majority of these loads could be shed first. The interruptible load policy may be a potential option if it proves to be beneficial for both the utility and customers. The fourth alternative, therefore, is to shed the interruptible loads when necessary (Alternative 4). In risk evaluation, the interruptible loads would be shed first to maintain system security (no overloads) and supply the firm loads. The minimum damage cost model given in Section 6.3.2 can guarantee that this curtailment principle is performed automatically. This alternative was included in the project only for study purposes since no formal load shedding agreement had been made between the utility and the industrial customers at that time.

6.4.3 Risk Evaluation

In the risk analysis, the following three annual indices are used:

1. EENS: Expected Energy Not Supplied (MWh/yr)
2. EFLC: Expected Frequency of Load Curtailments (occurrences/yr)
3. EDLC: Expected Duration of Load Curtailments (h/yr)

The annual indices are calculated considering an annual load duration curve. The economic analysis based on risk cost assessment requires utilization of annual indices. The coefficient of variance for the EENS is set at 0.05 in the Monte Carlo simulation.

The annual EENS indices of the base case (A0) and four alternatives for the 1995 to 2003 load levels are listed in Table 6.1. The comparison of the results is

Table 6.1   Annual EENS(f) indices for the base case and four alternatives (MWh/yr)

Load Level Year	A0	A1	A2	A3	A4
1995	370.12	262.60	315.40	338.18 (339.78)	45.14 (61.39)
1997	435.86	287.09	351.01	374.66 (377.72)	56.35 (75.33)
1999	547.72	332.67	398.24	426.39 (432.46)	73.20 (96.33)
2001	768.19	414.25	471.31	524.66 (536.83)	105.03 (138.19)
2003	1171.37	558.41	589.21	707.17 (730.38)	154.87 (205.70)

also shown in Figure 6.3. A1 denotes the 2L35 line addition alternative, A2 the "cut-and tie" alternative, A3 the operational manipulation alternative, and A4 the interruptible industrial load alternative. Note that the indices are only associated with the curtailments of the firm loads. In other words, the load shedding associated with the interruptible loads is not included in the indices for A4. To characterize this, the symbol of EENS(f) has been used in the captions of Table 6.1 and Figure 6.3. Alternatives A3 and A4 need the use of an automatic control or load shedding scheme that may not be 100% reliable. The numbers not in the brackets in Table 6.1 are the annual EENS indices obtained by assuming the 100% success probability of the automatic control or load shedding scheme, whereas the numbers in the brackets correspond to the 95% success probability.

It can be seen that the interruptible load alternative provides the best reliability improvement. It requires, however, a compensation to the customers. The 2L35 addition offers the second-best reliability improvement. This alternative requires a

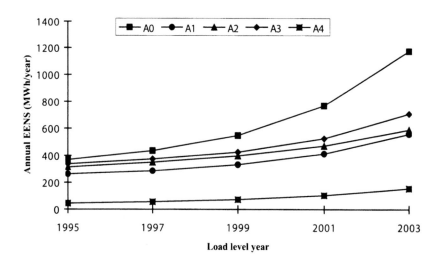

Figure 6.3   Annual EENS(f) indices for the base case and four alternatives.

large investment. The "cut-and-tie" alternative results in a slightly higher overall risk level than the 2L35 addition but with a much lower investment. The operational manipulation alternative produces a reliability improvement that is fairly close to that provided by the "cut and tie" option. This alternative requires the revision of the operation guideline and creates the complexity in operation.

The interruptible load alternative leads to the smallest annual EENS(f) index because the load shedding associated with the interruptible loads is not counted in the EENS(f) index. This exclusion is reasonable since the load curtailments reflected in the EENS(f) index are by nature different from the load shedding due to the interruptible loads. Inconveniences or even limited damages, however, would be brought to the interruptible customers due to implementation of the interruptible load policy. It is necessary to utilize a set of indices to indicate this kind of consequence. In order to distinguish the indices due to the interruptible loads from those due to the firm loads, a postfix (I) is added to the symbols of the indices. The annual EENS(I), EFLC(I) and EDLC(I) indices listed in Tables 6.2 and 6.3 provide the information about how often, how long, and how seriously the industrial customers experience the load shedding due to the interruptible policy on the average basis. Table 6.2 shows the indices of eight industrial customers as an interruptible load group and Table 6.3 the individual indices for each bus. Note that the EFLC(I) and EDLC(I) indices for the whole group are not the sum of the corresponding bus indices. All the indices are based on the condition of the 100% success probability for the load shedding scheme. The frequency and duration indices due to the load shedding for the industrial customers as a group show relatively small increases, whereas the expected energy not supplied shows relatively large increases as the load level grows with the years. This indicates that the number of system contingency states causing overloading is just slightly impacted but the amount of overloading is significantly raised by the load level growth in this example.

6.4.4 Overall Economic Analysis

6.4.4.1 Approach. The above risk indices provide a comparison for the four alternatives. This comparison is meaningful but insufficient since the investment cost cannot be directly compared with the EENS index. Moreover, the different effects

Table 6.2 Annual EENS(I), EFLC(I) and EDLC(I) indices for the eight interruptible industrial customers as a group

Load Level Year	EENS(I) (MWh/yr)	EFLC(I) (occurrences/yr)	EDLC(I) (h/yr)
1995	324.98	4.97	101.93
1997	379.51	5.10	103.37
1999	474.52	5.21	107.23
2001	663.16	5.62	108.74
2003	1016.50	5.71	118.19

Table 6.3   Annual EENS(I), EFLC(I) and EDLC(I) indices for each interruptible customer

Index	Bus No.	Load Level Year				
		1995	1997	1999	2001	2003
EENS(I)	631&632	113.41	141.75	179.93	265.63	425.61
(MWh/yr)	699	146.70	153.76	162.31	176.90	184.80
	694	30.81	36.45	46.50	53.42	88.87
	690	17.44	20.77	29.65	49.12	97.40
	630	7.88	10.07	21.53	43.72	111.48
	689	3.63	6.91	14.33	19.13	43.35
	623	3.16	5.81	12.00	32.97	39.13
	789	1.95	3.99	8.27	22.27	25.86
EFLC(I)	631&632	0.79	0.80	0.84	0.85	0.94
(occurrences/yr)	699	1.56	1.57	1.58	1.62	1.63
	694	1.99	2.00	2.02	2.03	2.06
	690	0.61	0.61	0.67	0.80	0.84
	630	0.01	0.02	0.11	0.37	0.40
	689	0.08	0.20	0.21	0.38	0.41
	623	0.01	0.02	0.03	0.08	0.10
	789	0.01	0.02	0.03	0.08	0.08
EDLC(I)	631&632	10.77	11.83	15.42	16.03	24.62
(h/yr)	699	86.37	86.90	87.60	91.45	92.07
	694	4.91	5.43	6.48	7.10	10.69
	690	1.66	2.01	2.80	4.29	8.58
	630	0.53	0.61	2.10	3.15	6.66
	689	0.88	1.58	2.63	3.50	7.27
	623	0.61	1.14	2.10	5.78	7.18
	789	0.53	1.14	2.10	5.78	6.31

of load curtailments due to the firm and interruptible customers are not recognized on a value basis. It is therefore necessary to conduct a further economic analysis based on the risk cost assessment.

As mentioned in Section 6.2.1, the best alternative should achieve the minimum total cost, which is the sum of the investment, operating, and risk costs in a given period. According to Equations (6.2) to (6.4), the present value of the total investment cost (*TI*) for a period of *m* years can be calculated by

$$TI = V \frac{i(1 + i)^n}{(1 + i)^n - 1} \sum_{j=1}^{m} \frac{1}{(1 + i)^{j-1}} \tag{6.13}$$

where V is the capital cost of the added system facility in the starting year. In this example, the capital is invested only in the first year. This leads to the same annual equivalent investment in the period of *m* years. Otherwise, a cash flow to capture multiple stage investments must be created (see Section 6.5).

Using Equations (6.4) and (6.12), the present value of the total risk cost due to the system unreliability (TR) in the same period can be given by

$$TR = \sum_{j=1}^{m} \frac{d_s \cdot EENS(I)_j + d_c \cdot EENS(f)_j}{(1 + i)^{j-1}} \tag{6.14}$$

where d_s and d_c are the unit damage costs (in \$/megawatt hour) for the interruptible and firm loads, respectively, which are the same as defined in the risk cost model given in Section 6.3.2; and $EENS(I)_j$ and $EENS(f)_j$ are the EENS indices (in MWh/yr) due to, respectively, the interruptible and firm loads for year j. If an alternative is not associated with the interruptible load policy, the term $d_s \cdot EENS(I)_j$ does not exist.

The operating cost for transmission systems includes the network loss cost and other expenses. Generally, the present value of the total operating cost (TO) for the same period can be obtained by directly using Equation (6.4):

$$TO = \sum_{j=1}^{m} \frac{O_j}{(1 + i)^{j-1}} \tag{6.15}$$

where O_j is the annual operating cost for year j and i is the discount rate. In this example, the operating cost is excluded from the total cost for comparison since it is basically the same for the four alternatives.

6.4.4.2   *Data.*   In the economic analysis, the following data were used:

1. The economic life for any investment, including the line 2L35 addition in Alternative 1, the "cut and tie" in Alternative 2, and the automatic control and monitoring device in Alternative 3, is assumed to be 45 years. The total capital costs for the line 2L35 addition, the "cut-and-tie" scheme and the automatic control and monitoring device in 1997 dollars are estimated as \$26,431,543, \$140,000, and \$100,000, respectively. The discount rate is 10%.

2. The period considered is the 7 years from 1997 to 2003. No matter which alternative was applied, no other reinforcement would be considered before 2003.

3. Alternative 4 was included only for study purposes since there was no agreement yet between the utility and the interruptible customers. It is assumed that the damage cost caused by the interruptible industrial loads is only the revenue reduction to the utility and no compensation to the industrial customers is considered. The average tariff rate (\$60/MWh) is used as d_s.

4. The damage cost caused by the supply interruption of the firm loads should include various direct and indirect factors such as the lost revenue to the utility, customer's direct damage, and intangible societal and environmental aspects. The average unit damage cost d_c is estimated to be \$6300/MWh according to the customer survey and the capital planning guidelines of the utility.

5. The EENS indices of the four alternatives for 1997 to 2003, which are obtained from the risk evaluation, are summarized in Table 6.4. The interruptible load alternative (A4) is associated with both the EENS(f) and EENS(I), whereas the other three alternatives are only associated with the EENS(f). The indices for A3 and A4 correspond to the 95% success probability of the automatic control or load shedding scheme that is needed to implement the alternatives. Note that the EENS(I) indices in Table 6.2 correspond to the 100% success probability of the load shedding scheme.

6.4.4.3 *Results.* The present values (in 1997 dollars) of the total costs (investment and risk costs) of the four alternatives for the seven years considered (from 1997 to 2003) are:

$$TC_{A1} = 14.35 + 12.65 = 27.00 \text{ M\$}$$
$$TC_{A2} = 0.08 + 14.57 = 14.65 \text{ M\$}$$
$$TC_{A3} = 0.05 + 16.55 = 16.60 \text{ M\$}$$
$$TC_{A4} = 0.0 + 4.14 = 4.14 \text{ M\$}$$

It can be seen from the overall economic analysis that the addition of 2L35 is the highest total cost alternative. The present values of the total costs for the local system reconfiguration (the "cut and tie" in the 69 kV subsystem) and operational manipulation alternatives are, respectively, about 54% and 61% of that required by the 2L35 addition. As mentioned earlier, the 69 kV subsystem reconfiguration alternative produces the system reliability level close to that provided by the 2L35 addition but requires a much lower investment.

Alternative 4 is the interruptible load scheme. The total cost of A4 is the damage cost caused by the firm load curtailments plus the revenue reduction due to shedding the interruptible industrial loads. It should be noted that the interruptible load alternative is very sensitive to the compensation policy. This factor was not included in the present value of the total cost because no agreement between the utility and the industrial customers was available in the actual project.

Table 6.4   Annual EENS indices for the four alternatives (MWh/yr)

Load Level Year	A1 EENS(f)	A2 EENS(f)	A3 EENS(f)	A4 EENS(f)	A4 EENS(I)
1997	287.09	351.01	377.72	75.33	360.53
1998	309.22	364.63	401.83	84.11	399.56
1999	332.67	398.24	432.46	96.33	451.39
2000	365.66	424.81	480.87	113.36	547.32
2001	414.25	471.31	536.83	138.19	630.00
2002	471.33	526.30	623.01	162.51	777.63
2003	558.41	589.21	730.38	205.70	965.67

6.4.5   Summary

The example given in this section shows an application of transmission risk evaluation in an actual planning project of a utility. The four alternatives for the south metro system development planning are evaluated using the presented method. The four alternatives are representative and consist of line addition, system reconfiguration, operational manipulation, and interruptible load management. The results indicate that the 230 kV line addition option is not overall cost effective and therefore should be avoided. This allows a major capital expenditure of $26.4 million (1997 dollars) to be saved. The local reconfiguration of the 69 kV subsystem could provide basically the same reliability level as the 2L35 line addition with the much lower investment and it was, therefore, finally recommended in the planning decision.

The interruptible load alternative was included only for study purposes due to its premature implementation. The compensation to the industrial customers could not be determined because there was no agreement between the utility and the customers at that time. It can be seen, however, that a load side management alternative like the interruptible policy can also be evaluated and compared with the others using the probabilistic planning method.

6.5   EXAMPLE 2: APPLYING DIFFERENT PLANNING CRITERIA [32, 63]

The second example is also an actual project, which was undertaken in 1996. The purpose of giving this example here is to provide a wider insight into applying the probabilistic planning concept while emphasizing the following:

- Two different planning criteria
- Reinforcement sequences
- Multiple stage investments

6.5.1   System and Planning Alternatives

Figure 6.4 shows the north metro 500 kV/230 kV/69 kV system of the utility. As the load level in the north metro region grows, overloading would occur on the circuit 2L40 during the peak load period in some single-contingency cases, particularly when either circuit 2L32 or 2L52 is out of service. The task of the development planning is to solve this problem by selecting a reliable and economic reinforcement alternative.

The traditional load flow and contingency analysis studies were conducted and the following three alternatives were worked out to solve the overloading problem that would appear with the load growth.

1. Uprating of the circuit 2L40. The uprating is done in the three stages: the north segment in 1998, the south segment in 2001, and the segment crossing a river in 2004.

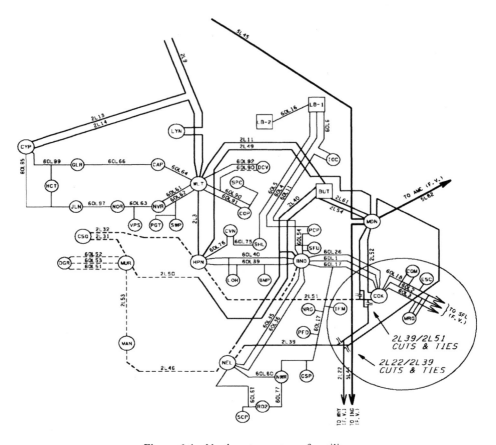

Figure 6.4   North metro system of a utility.

2. Cuts and ties of the circuits 2L22 and 2L39. As shown in Figure 6.4, the line 2L22 from MDN to WYH and the line 2L39 from NEL to COK are cut and tied into the line from NEL to MDN and the line from COK to WYH.

3. Cuts and ties of the circuits 2L39 and 2L51. The line 2L39 from NEL to COK and the circuit 2L51 from BND to MDN are cut and tied into the underground cable from BND to COK and the line from NEL to MDN.

Uprating 2L40 can alleviate the overloading problem in the long term but requires a high capital investment. The two alternatives of cuts and ties require a low capital cost but only solve the overloading problem on 2L40 for a relatively short time (before 2002). The purpose of the study was to evaluate the risks of the base system and three alternatives and compare the total cost efficiency from a probabilistic cost analysis viewpoint.

6.5.2 Study Conditions and Data

6.5.2.1 Study Conditions

1. To highlight the north metro region, only the north metro area system is included in the evaluation. This is the reduced load flow equivalence system that is obtained from the full base case of the utility system. The size of the north metro system is 114 buses and 173 lines/transformers.

2. According to the generation planning, a local generating station (BUT) that is located in the north metro area may be upgraded. This would produce significant impacts on the power flow distribution in the north metro area. Therefore, two scenarios are considered: (a) No real power is outputted at the BUT station and this represents the situation before the BUT upgrade. (b) The BUT upgrade project is taken into consideration and its real power output is assumed to be 900 MW before 2004 (inclusive) and 1400 MW after 2004. The base system and three transmission reinforcement alternatives are evaluated with each of the generation scenarios.

3. The time frame in the study is the 10 years from 1998 to 2007. The system load curve for 1995 is used to model the annual load profile. It is assumed that each year has 2% load growth and follows the same load duration curve shape. This means that 8760 hourly loads have the same proportional increase as the peak load point.

4. The forced outages of all the lines and transformers in the north metro area are included. Since the major sources causing the overloading on the circuit 2L40 are the outages of 2L32 and 2L52, the planned (maintenance) outages of these two circuits are also considered in the study.

6.5.2.2 Data

1. Network data. The network data is taken from the power flow base case.

2. Outage data. The outage data of the lines and transformers are based on the utility's historical records. The average forced outage data are shown in Tables 6.5, 6.6, and 6.7. The planned outage data are 2.8 and 2.0 maintenance

Table 6.5 Average forced outage data for transmission lines

Voltage (kV)	Line-Related		Terminal-Related	
	Frequency (occurrences/yr/100 km)	Repair time (h/occurrence)	Frequency (occurrences/yr/term.)	Repair time (h/occurrence)
60	7.329	4.38	0.091	9.51
230	0.726	10.71	0.107	13.96
500	0.598	8.19	0.387	7.42

Table 6.6 Average forced outage data for underground cables

	Cable-Related		Terminal-Related	
Voltage (kV)	Frequency (occurrences/yr/100 km)	Repair time (h/occurrence)	Frequency (occurrences/yr/term.)	Repair time (h/occurrence)
60	2.413	0.86	0.018	0.10
230	1.222	221.64	0.222	0.06

Table 6.7 Average forced outage data for transformers

	Component-Related		Terminal-Related	
Voltage (kV)	Frequency (occurrences/yr)	Repair time (h/occurrence)	Frequency (occurrences/yr)	Repair time (h/occurrence)
60	0.019	321.12	0.022	24.43
230	0.050	91.59	0.074	20.39
500	0.032	342.86	0.070	28.07

outages/year for 2L32 and 2L52, respectively. The average planned outage duration is estimated to be 1 day per outage.

3. Load curve data. The 8760 hourly load curve records are used and transformed into a multiple step load duration model.
4. Economic analysis data. The following economic analysis data are used in the study:

   (a) Unit interruption cost: $60/MWh
   (b) Discount rate: 8%
   (c) Economic life of investment: 45 years
   (d) Capital investment costs:

Existing system:	$0
2L40 uprating:	$2405 k in 1998
	$3291 k in 2001
	$2000 k in 2004
2L22/2L39 cuts and ties:	$1200 k in 1998
2L39/2L51 cuts and ties:	$1200 k in 1998

6.5.3 Risk and Risk Cost Evaluation

The following two risk indices are used:

1. EENS—Expected Energy Not Supplied (MWh/yr)
2. EDC—Expected Damage Cost (k$/yr)

The EDC index is calculated based on $60/MWh, which is the average electricity rate of the utility. Therefore, the EDC index in the study is the "real damage cost" to the utility rather than the social cost. This is a conservative assumption in risk cost evaluation.

The EENS and EDC indices of the base system and three alternatives from 1998 to 2007 for the first scenario (no generation output at the BUT station) are shown in Tables 6.8 and 6.9. Those for the second scenario (900 MW output before 2004 inclusive and 1400 MW after 2004 at the BUT station) are shown in Tables 6.10 and 6.11.

It can be seen from the results that the 2L40 uprating or the 2L39/2L51 cuts and ties would lead to better system reliability compared to the base system. The 2L22/2L39 cuts and ties, however, would create an overall system risk level much worse than the base system configuration, although they could reduce the loading level on 2L40 in some single contingencies. Another observation obtained from the

Table 6.8 EENS (MWh/yr) of the base system and three alternatives for Scenario 1 (without generation output at BUT station)

Year	Base	2L40 Uprating	2L22/39	2L39/51
1998	1032	695	12270	841
1999	1272	822	13209	945
2000	1585	956	14152	1053
2001	2040	1126	15129	1179
2002	2604	1406	16197	1384
2003	3249	1698	17273	1601
2004	4156	2071	18359	1922
2005	5203	2588	19531	2319
2006	6289	3151	20765	2753
2007	7613	3765	22047	3444

Table 6.9 EDC (k$/yr) of the base system and three alternatives for Scenario 1 (without generation output at BUT station)

Year	Base	2L40 Uprating	2L22/39	2L39/51
1998	62	42	736	51
1999	76	49	793	57
2000	95	57	849	63
2001	122	68	908	71
2002	156	84	972	83
2003	195	102	1036	96
2004	249	124	1102	115
2005	312	155	1172	139
2006	377	189	1246	165
2007	457	226	1323	207

Table 6.10 EENS (MWh/yr) of the base system and three alternatives for Scenario 2 (with generation output at BUT station)

Year	Base 2L40 Uprating	2L22/39	2L39/51	
1998	775	433	12016	626
1999	931	473	12881	657
2000	1156	515	13750	693
2001	1494	563	14654	736
2002	1839	619	15582	785
2003	2259	682	16518	845
2004	2873	749	17464	925
2005	3449	792	18403	1049
2006	4084	888	19429	1167
2007	4958	999	20504	1411

Table 6.11 EDC (k$/yr) of the base system and three alternatives for Scenario 2 (with generation output at BUT station)

Year	Base	2L40 Uprating	2L22/39	2L39/51
1998	47	26	721	38
1999	56	28	773	40
2000	69	31	825	42
2001	90	34	879	44
2002	110	37	935	47
2003	136	41	991	51
2004	172	45	1048	56
2005	207	48	1104	63
2006	245	53	1166	70
2007	298	60	1230	85

results is that the BUT station upgrade project would improve the reliability in the north metro system. Obviously, this is due to the fact that the BUT upgrade would bring local generation capacities into the regional system so that the loading on some lines would be decreased, particularly on those lines connected to the main power supply sources outside the region.

6.5.4 Overall Economic Analysis

In this application, the overall economic analysis is conducted using two criteria: (1) the performance-based cost efficiency criterion, and (2) the combined criterion of single-contingency and performance-based cost efficiency. In either case, the total costs are calculated for the comparison between the different alternatives. The transmission system operation cost is due to network losses. The losses are evaluated using the power-flow program that can consider a load curve. The results indicate that the network losses for the base system and three alternatives are basically the same. Therefore, the operation cost is excluded from the total cost and only the investment and risk costs are considered. The expected damage cost for each year is calculated using Equation (6.12). Note that the term $d_s \cdot EENS(I)$ is zero in this example as no alternative is associated with the interruptible policy. The calculation for the cash flow of the investment cost for the 2L40 uprating alternative is slightly different since it has to capture multiple-stage investments.

6.5.4.1 *Cash Flows in Multiple-Stage Investments.* Equation (6.13) is
applicable to a single capital investment in the starting year, such as the case in Example 1. When multiple-stage investments are considered, the basic concepts presented in Sections 6.2.2 and 6.2.3 are still valid. The key is determination of the total cash flow for the multiple-stage investments. The cash flow for each individual investment can be obtained using Equations (6.2) and (6.3). Each cash flow has a time length that is equal to the economic life of the investment, with each year having equal annual equivalent capital. The cash flows corresponding to different investment stages show a time shift between them and can be superposed to obtain the total cash flow for the multiple-stage investments. This concept is illustrated in Figure 6.5. The investments 1, 2, and 3 are assumed to be in place in years T_1, T_2, and T_3, with annual equivalent capitals A_1, A_2, and A_3, respectively. The planning span is from T_1 to T_4 (the dashed line). The total cash flow for the three-stage investments within the planning span should be calculated in such a way that the annual capital between T_1 and T_2 is A_1, the one between T_2 and T_3 is $A_1 + A_2$, and the one between T_3 and T_4 is $A_1 + A_2 + A_3$. Once the total cash flow is obtained, Equation (6.4) is used to calculate the present value.

6.5.4.2 *Performance-Based Cost Efficiency Criterion.* Under this crite-
rion, the $N-1$ principle is not treated as an imperative requirement. The single contingencies are considered as a part of the events impacting the overall system risk. The effects of the single contingencies combined with their probabilities of occurrence are included in the risk cost component.

The present values (1998 dollars) of the investment, risk and total costs for the base system and three alternatives in the period from 1998 to 2007 are shown in Ta-

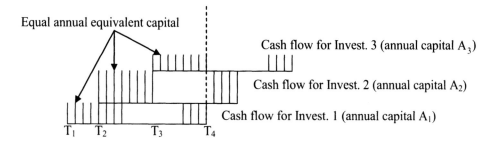

Figure 6.5 Cash flows for multiple-stage investments.

Table 6.12   Investment, risk, and total costs (PV in 1998 dollars) of the base system and three alternatives for 1998 to 2007 (without generation output at BUT station)

Cost	Base System (k$)	2L 40 Uprating (k$)	2L22/39 (k$)	2L39/51 (k$)
Investment	0	3025	718	718
Risk	1330	705	7052	687
Total	1330	3730	7770	1405

bles 6.12 and 6.13 for the two BUT generation scenarios, respectively. It can be seen that the base system requires the lowest total cost followed by the alternative of 2L39/2L51 cuts and ties. The alternative of 2L22/2L39 cuts and ties leads to the highest total cost mainly due to the risk cost component caused by the overall system unreliability, although it can reduce the loading level on 2L40. The 2L40 uprating has the low risk cost but needs high capital investment, resulting in total cost inefficiency. The results suggest that the base system configuration is the best one from the performance-based cost efficiency perspective if meeting the single-contingency principle is not a requirement. Compared to the base system, the alternative of 2L39/2L51 cuts and ties could meet the single-contingency criterion before 2002 and the price paid to obtain this is the slightly increased total cost that ranges from $75 k (1405 – 1330) to $173 k (1086 – 913) for the scenarios without and with the output at the BUT generating station.

6.5.4.3   Combined Criterion of Single Contingency and Cost Efficiency.

The north metro area is one of the most important service regions in the utility. The single-contingency criterion should be a mandate in the decision-making process. A combined criterion of single-contingency and performance-based cost efficiency is also applied in the study. The base system cannot meet the single-contingency requirement. The 2L40 uprating can meet the single-contingency principle but requires the high capital investment. The two cut and tie alternatives can meet the single-contingency requirement only before 2002 and the 2L39/2L51 alternative has the much better reliability than the 2L22/2L39 alternative. Based on these facts, the following two sequences that satisfy the single-contingency requirement are compared using the performance-based cost efficiency criterion:

Table 6.13   Investment, risk, and total costs (PV in 1998 dollars) of the base system and three alternatives for 1998 to 2007 (with generation output at BUT station)

Cost	Base System (k$)	2L40 Uprating (k$)	2L22/39 (k$)	2L39/51 (k$)
Investment	0	3025	718	718
Risk	913	276	6753	368
Total	913	3301	7471	1086

Sequence 1: 2L40 uprating. This is done in the three steps: (1) the north segment upgrade in 1998; (2) the south segment upgrade in 2001; and (3) the crossing river segment upgrade in 2004.

Sequence 2: 2L39/2L51 cuts and ties and deferred 2L40 uprating. This includes the following stages: (1) 2L39/2L51 cuts and ties in 1998; (2) 2L40 uprating deferred to start the first step in 2002, the second step in 2005, and the third step in 2008.

The present values (1998 dollars) of the investment, risk, and total costs for the two sequences in the period from 1998 to 2007 are shown in Tables 6.14 and 6.15 for the two generation scenarios, respectively. It can be seen that Sequence 2 can save about $1300 k of capital investment through deferring the 2L40 uprating by 4 years and that both the sequences meet the single-contingency requirement. Sequence 2 requires the lower total cost and provides almost the same risk level.

6.5.5 Summary

In this application example, the probabilistic planning results indicate that if the single-contingency requirement is not a must, the base system (i.e., doing nothing) is the best option in terms of total cost efficiency, including the investment and risk costs. Based on the combined criterion of single-contingency and performance-based cost efficiency, Sequence 2 (i.e., 2L39/2L51 cuts and ties in 1998 and 2L40 uprating in three steps with the first one in 2002) was recommended in the actual project. Compared to starting the 2L40 uprating in 1998, Sequence 2 would save

Table 6.14 Investment, risk, and total costs (PV in 1998 dollars) of the two sequences for 1998 to 2007 (without generation output at BUT station)

Cost	Sequence 1 (k$)	Sequence 2 (k$)	Difference (k$)
Investment	3025	1730	1295
Risk	705	687	18
Total	3730	2417	1313

Table 6.15 Investment, risk, and total costs (PV in 1998 dollars) of the two sequences for 1998 to 2007 (with generation output at BUT station)

Cost	Sequence 1 (k$)	Sequence 2 (k$)	Difference (k$)
Investment	3025	1730	1295
Risk	276	357	−81
Total	3301	2087	1214

about $1.3 million of capital cost while maintaining almost the same reliability level and meeting the single-contingency criterion. From the viewpoint of the total cost efficiency and system reliability, the option of 2L22/2L39 cuts and ties should be avoided, although it could meet the single-contingency requirement before 2002. The weakness of this option could not have been identified if the risk-based probabilistic planning were not conducted.

6.6   CONCLUSIONS

The application of risk evaluation in transmission development planning is often called probabilistic planning. This chapter discusses its concept, methods, and procedures. The quantified risk assessment and the overall economic analysis are two basic aspects. Although the investment and operating costs are normally included in utility system planning, inclusion of the risk cost adds another dimension that enhances planning decisions.

Two actual examples in a utility have been given to demonstrate the details of the application. In transmission system development planning, a line addition is a usual consideration. A change in the local network configuration is also an important option. From an integrated planning point of view, on the other hand, noninvestment alternatives associated with operational manipulation and load side management may be cost-effective and competitive. Therefore, a risk evaluation method should be able to assess all possible alternatives, including both investment-type and noninvestment-type reinforcements. This has been illustrated in the first example. In many cases, planning alternatives are associated with reinforcement sequences and multiple-stage investments. It is desirable for utilities to apply the probabilistic cost efficiency criterion and keep the deterministic $N - 1$ principle that has been used for years. There is no conflict between the single-contingency and probabilistic performance-based criteria. The combined criterion can be applied when the single-contingency principle is still a mandatory requirement. These aspects have been discussed in the second example.

Probabilistic planning is a fundamental and powerful tool that can provide utilities with significant economic benefits while maintaining an acceptable system risk level.

CHAPTER 7

APPLICATION OF RISK EVALUATION TO TRANSMISSION OPERATION PLANNING

7.1 INTRODUCTION

The essential operation planning function in a control center is to select secure, reliable, and economic operation modes under different operation conditions. The technical analysis in operations has for years been usually based on the deterministic principle. Compared to transmission development planning, a difference is that in addition to the single-contingency $(N - 1)$ criterion, a few double- or even triple-contingency events associated with critical equipment are also studied in operation planning. This is partly because system operation has to deal with the situation in which a component may fail when other components are out of service for maintenance and partly because simultaneous failures of multiple components can actually happen in real life.

The probabilistic risk evaluation of transmission operation modes should be addressed as a part of the operation planning function for the following reasons:

- Any system, even if it meets the requirement of the single-contingency criterion, still has an operational risk due to higher failure levels. Many power outages and blackouts are caused by multiple equipment failures.
- There often exist several operation modes that all meet the deterministic operation criteria. It is necessary to identify the lowest-risk operation mode to assure overall system reliability. This is particularly essential in the deregulated and competitive environment of the power industry.
- The deterministic operation criteria are based on worst-case studies, including

the peak load and extreme operation conditions. It is impractical to cover a huge number of multiple component failure events using a deterministic approach. Risk evaluation provides the risk index that combines consequences of all possible failure events with probabilities of their occurrence. This is obviously a valuable enhancement to the deterministic operation analysis technique.

Risk assessment of transmission systems has been considered in system development planning at some utilities but has not been sufficiently addressed in operation planning so far. A fundamental difference between system development planning and operation planning is that the former is normally associated with system reinforcements in a long time frame (several to more than 10 years), whereas the latter is associated with operational measures in a short term (less than one year). As seen in Section 6.4, an operational manipulation can be an option in the development planning alternatives. However, such an alternative generally requires the permanent modification of an operation rule, not an operational measure in a short period. There is no equipment addition in operation planning. The operation measures include load transfers, generation pattern changes, temporary network reconfigurations, and switching actions. This chapter discusses the risk evaluation of operation modes, which requires some new features in the assessment method. The concept and method of operation risk assessment are described in Sections 7.2 and 7.3, respectively. Two application examples are given in Sections 7.4 and 7.5 to demonstrate the particulars of the presented method.

7.2 CONCEPT OF RISK EVALUATION IN OPERATION PLANNING [26]

The basic philosophy of transmission system risk assessment for system development planning and operation planning is quite similar—high level failure events associated with multiple-component outages must be evaluated, including their consequences and probabilities of occurrence. On the other hand, there are some special requirements in operation risk evaluation and these are related to the simulation of operation measures, including load transfers, generation pattern changes, network reconfigurations, and equipment switching. Each of the measures causes an incremental change (positive or negative) in system risk. The operation measures are temporary and are generally not considered in the risk evaluation for system development planning.

A load transfer only changes the location of loads but does not change the total load level in a system. The transfer could move a part of risk from one area to another. It is interesting that the decrease of risk in one area where a load is moved out is not equal to the increase of risk in another area where the load is moved in. In some cases, the transfer reduces the operation risk of the overall system. In the risk evaluation of system development planning, a constant load profile at each bus is normally assumed. Although an increase of the annual peak load with years is considered in development planning, this is completely different from the load transfer concept in operation.

Changing generation patterns is a common action in system operation. The change may be needed because of either operation security considerations such as generation rejections in some system states, or time-dependent generation source limits such as reservoir dispatch and/or maintenance requirements. Generally, the total generation is kept unchanged for a system steady state since the total load is the same. The pattern change means adjustments in generation output capacity at different generators. In the risk evaluation of system development planning, a constant generation capacity (the maximum output limit) at each generator is assumed. Note that a derated state of generators in risk evaluation is a different concept from the generation pattern change.

Network reconfigurations and equipment switching actions in operation are associated with the temporary modification of network topology by opening or closing breakers. Obviously, this will change power-flow distributions and thus system risk. The network reconfigurations are usually related to a relatively long period (several days to several months), whereas the switching is system state dependent and only lasts until a failed component recovers or the system gets into another operation condition. In the risk evaluation of system development planning, no switching action is considered. The main benefit of the network reconfigurations and equipment switching is the possibility to move a part of risk from one area to another or from one voltage level to another with reduced total risk of the whole system.

An important phenomenon associated with the network reconfigurations and equipment switching is the so-called noncoherence in transmission system reliability [1, 26]. The noncoherence refers to the fact that if one more component goes out of service, system reliability will at least not be deteriorated or may even become better. This phenomenon cannot occur in generation system risk evaluation but can happen in transmission systems. It can be explained using the four-bus system shown in Figure 7.1. The two generators at buses (1) and (4) supply the two loads at buses (2) and (3). Each load is 50 MW + 15 MVAR. Line 1 and line 3 have impedance of 0.008 + j0.034 p.u., the impedance of line 2 is 0.001 + j0.003 p.u., and the impedance of line 4 is 0.005 + j0.015 p.u.. All the lines have a rating of 55 MVA.

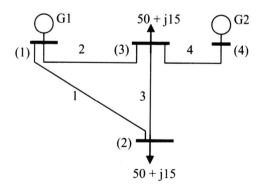

Figure 7.1 Four-bus system (normal state).

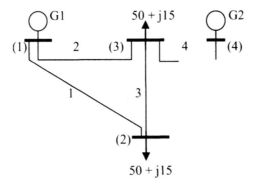

Figure 7.2   Four-bus system (line 4 outage).

The power-flow results show that there is no overload in the normal system state. When line 4 fails, the power from G2 is interrupted and both the loads are supplied by G1 (see Figure 7.2). In this outage state, line 2 is overloaded due to the power flow redistributions imposed by the Kirchhoff's laws. In order to alleviate the over-loading, some load curtailments at either load bus have to be performed. However, if we intentionally open line 3 (see Figure 7.3), the overloading problem on line 2 will disappear. The line flows at the "from bus" side for the three cases are given in Table 7.1. Note that there are losses on the lines so that the power flows at the two ends of each line are different.

This simple example provides the basic concept of noncoherence—losing lines 3 and 4 will create a more reliable system state than will just losing line 4. It is impor-tant to appreciate that a noncoherence event is dependent on system states. The in-tentional opening leading to a more reliable system state does not mean that the opened component is surplus and should be removed permanently. Obviously, line 3 in this example is needed in the normal and other contingency states to make the system more reliable. The noncoherence feature in transmission systems provides

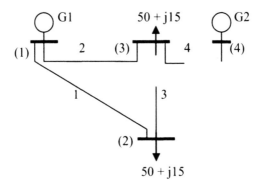

Figure 7.3   Four -bus system (line 4 outage and line 3 intentionally opened).

Table 7.1 Line flows corresponding to Figures 7.1, 7.2, and 7.3

Line	Bus to Bus	Line flow (MVA)	Rating (MVA)
Normal state			
1	(1) to (2)	27.36	55
2	(1) to (3)	25.30	55
3	(3) to (2)	25.07	55
4	(4) to (3)	52.20	55
Line 4 in outage			
1	(1) to (2)	29.61	55
2	(1) to (3)	75.11	55 (overload)
3	(3) to (2)	22.81	55
Line 4 in outage and line 3 opened			
1	(1) to (2)	52.64	55
2	(1) to (3)	52.25	55

the possibility for system operators to reduce system operation risk by changing operation modes or system configurations. This will be further demonstrated and explained in the application to a utility system given in Section 7.4. It should be stressed that the evaluation techniques of transmission system risk that are based on the assumption of coherence cannot be used in the risk evaluation associated with network reconfigurations and switching actions.

7.3 RISK EVALUATION METHOD [26, 28]

The Monte Carlo simulation based method for the risk evaluation of transmission operation modes includes the following steps:

1. Determine the operation modes using off-line power flow and contingency analysis, which may be associated with load transfers, generation pattern changes, and/or network reconfigurations.
2. Select the load curve representing a time period (weekly, monthly, or seasonal).
3. Select the up or down state of system components (lines, cables, transformers, generators, etc.) using Monte Carlo simulation.
4. If there is a need, conduct system-state-dependent switching actions, load transfers, and/or generation pattern changes based on the selected system state and operation criteria.
5. Perform automatic contingency analysis to identify if there is a system problem (such as overloading, isolated buses, or even islands).
6. If there is a system problem, the linear programming optimization model is utilized to reschedule generation outputs, alleviate line overloads, and avoid

load curtailments if possible, or minimize the total load curtailment if unavoidable. The mathematical expression of the optimization model is as follows:

$$\min \sum_{i \in ND} W_i C_i \tag{7.1}$$

subject to:

$$T(S^k) = A(S^k)(PG - PD(S^k) + C) \tag{7.2}$$

$$\sum_{i \in NG} PG_i + \sum_{i \in ND} C_i = \sum_{i \in ND} PD_i(S^k) \tag{7.3}$$

$$PG^{\min}(S^k) \leq PG \leq PG^{\max}(S^k) \tag{7.4}$$

$$0 \leq C \leq PD(S^k) \tag{7.5}$$

$$|T(S^k)| \leq T^{\max}(S^k) \tag{7.6}$$

where S^k denotes the system state for the kth forced failure event; $T(S^k)$ is the real power line flow vector for the system state S^k; $A(S^k)$ is the relation matrix between real power line flows and power injections for the system state S^k; PG is the generation output vector and PG_i is its element; $PD(S^k)$ is the bus load vector and $PD_i(S^k)$ is its element; C is the bus load curtailment vector and C_i is its element; $PG^{\max}(S^k)$ and $PG^{\min}(S^k)$ are the upper and lower limit vectors of generation variables, respectively; $T^{\max}(S^k)$ is the real power rating vector of lines; W_i is the weighting factor reflecting bus importance; ND is the set of load buses; and NG is the set of generation buses.

It is important to note that not only $T(S^k)$ and $A(S^k)$ but also $PD(S^k)$, $PG^{\max}(S^k)$, $PG^{\min}(S^k)$, and $T^{\max}(S^k)$ are system state dependent in the model.

7. Update the risk indices based on the C_i found from the model and the state probability from Monte Carlo sampling.

8. Repeat steps 3 to 7 until convergence is reached in the Monte Carlo simulation.

9. Go to step 2 for the load curve representing the next time period. This is repeated until all periods have been considered.

10. Go to step 1 for the next operation mode. This is repeated until all operation modes considered have been examined.

It can be seen from the above procedure and optimization model that the risk evaluation method used in operation planning has several new features compared to that used for system development planning. These mainly include:

- Since load transfers and generation patterns may be system state dependent, the bus loads and generation limits are the functions of individual system states in the optimization model.
- The dynamic ratings of circuits may need to be considered. The dynamic rating refers to the different rating values of a circuit in contingency system states. It depends on the flow on the circuit in a precontingency state and the duration of a contingency that relies on the repair time of failed components. To formulate the dynamic rating concept, the line limits T^{max} have been expressed as the function of a system state in the model.
- The Monte Carlo simulation must be able to handle the switching actions following an outage event and the possible noncoherence phenomenon. Any method based on the assumption of system coherence cannot be used.
- The risk evaluation in system development planning is generally on a yearly basis and, therefore, deals with an annual load curve. In operation planning, the evaluation must be flexible enough to handle a load curve of any time length such as a weekly, monthly, or seasonal load curve. This is because the time frame in operation planning ranges from a couple of days to several months. Also, there are many time-period-dependent factors. For instance, the generation patterns and line/transformer ratings are varied in different seasons and even a relatively short time period in the risk evaluation may cross a season. Equipment maintenance can take place at any point of time.
- The bus indices are important to quantify how much risk can be moved from one area to another for an operation mode. To make the bus indices realistic, the weighting factors (W_i) have to be specified properly based on the importance of substations. In the presented model, W_i only needs to be specified in accordance with their relative importance.

7.4 EXAMPLE 1: DETERMINING THE LOWEST-RISK OPERATION MODE [72, 73]

This example is an actual application of the presented method in operation planning at the control center of a utility.

7.4.1 System and Study Conditions

The application was performed on the 230 kV and 60 kV metro systems of the utility shown in Figure 7.4. It includes the north and south metro regions. The key branches and substations mentioned below have been highlighted with bold lines in the figure. The problem was that a critical underground cable 2L39 had to be replaced because of its end-of-life failure. The replacement project would take the 9 months from October 2000 to June 2001. During the replacement, the operation risk in the north metro region, including the downtown area of a big city, would be greatly increased due to the absence of 2L39. One task of the operation planning was to identify the lowest-risk operation mode and assure the reliability of the

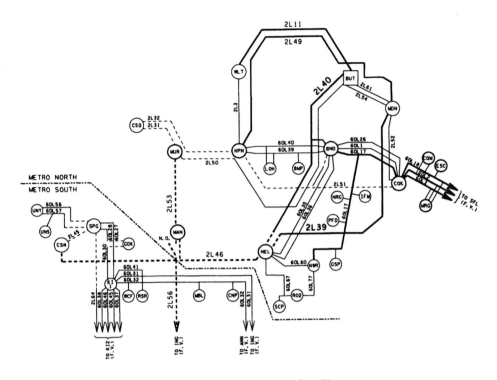

Figure 7.4 Metro system of a utility.

downtown area during this period. The EENS (expected energy not supplied) was used as an index to reflect the system risk.

The study conditions included:

- The time frame was 9 months. The months from October to March were assumed to be in the winter and those from April to June in the summer.
- The different load curves, line/transformer ratings, and local generation patterns for the winter and summer periods were considered.
- The dynamic ratings of cables were applied.
- The outage data of the lines, cables, and transformers were based on the failure statistics of the previous 15 years.
- The aging failures of the cables in the system were also considered.
- The terminal-related line outages were included.
- The dependent failures of some line sections were modeled.

7.4.2 Assessing Impacts of Load Transfer

In order to reduce the operation risk in the north metro system when 2L39 was out of service, two possible measures were identified using a contingency analysis:

1. The load at Substation MAN could be transferred to the south metro system and supplied from line 2L56. This load would be transferred back to the north metro system if 2L56 failed.

2. The 230/60 kV transformer T3 at Substation HPN could be closed from its normally open status.

The first measure, with assistance of the second one, would allow the system to still satisfy the single-contingency criterion. The first task of the risk evaluation was to answer the following questions: How much risk would be introduced in the south metro system due to the load transfer? Would this be acceptable? What would be the total risk for the whole metro system?

The EENS indices are given in Table 7.2. It can be seen that the risk in the north metro system with 2L39 out of service plus the load transfer is even lower than that in the normal state with 2L39 in service but without the load transfer. This indicates that the load transfer can offset the negative impact due to 2L39 being out of service on the north metro system. On the other hand, there is a risk increase in the south metro system due to a partial risk shift resulting from the load transfer. This risk increase, however, is acceptable because the south metro system can still meet the single-contingency criterion. Particularly, the risk evaluation indicates that the total risk in the whole metro area is only marginally increased. The load transfer is a successful measure to control the north metro system risk while 2L39 is out of service. The disadvantage is the complexity in operation due to the load switching back, which is needed if 2L56 fails. The operators are required to perform closer monitoring and act faster when necessary. However, this price has to be paid to assure the reliability of the overall system.

7.4.3 Comparing Different Reconfigurations

The major impact due to 2L39 being out of service is the overloading on cable 2L40 that parallels 2L39. After the load transfer at Substation MAN to the south metro system, the overloading problem is resolved but the loading level on 2L40 is still high. Cable 2L40 is also unreliable due to its aging and bad failure records in the past. The high loading level would increase the failure probability. Two options are identified to further reduce the flow on 2L40:

1. The two 230/60 kV transformers T4 and T5 at Substation NEL are opened so that 2L40 will only supply the loads at NEL while it originally also supplies the 60 kV system connected to NEL.

2. After the load transfer at MAN, the cables 2L46 and 2L53 become one circuit. If this circuit is opened, more power will flow westward through the north path (2L11 and 2L49), resulting in a reduction of the loading on 2L40.

The second task of the risk evaluation was to answer the questions: Which option would be better in terms of operation risk? Would there be any additional risk caused by either of the two options?

Table 7.2 EENS (MWh) for the cases with 2L39 and without 2L39 but with load transfer

	North system		South system		
	Oct.–Mar.	Apr.–Jun.	Oct.–Mar.	Apr.–Jun.	Total
With 2L39	2603	1206	401	253	4463
No 2L39 but with the load transfer	1939	887	1023	744	4593

The EENS indices for the three cases are shown in Table 7.3. The first case is the base case—2L39 out of service and the load transfer at MAN to the south metro system. The second one is the base case plus option 1—the transformers T4 and T5 at NEL are opened. The third one is the base case plus option 2—2L46/53 is opened. Note that the EENS indices for the south metro system are the same for the three cases. It is interesting to note that option 1 would lead to higher operation risk in the north metro system. This is because the risk decrease due to the reduced flow on 2L40 could not offset the risk increase in the 60 kV system connected to Substation NEL. Option 2 of causing an increased loading level on the north path (the overhead lines) would result in higher risk during the summer months but lower risk during the winter months in the north metro system, with the overall risk level close to that of the base case. In fact, circuit 2L46/53 only plays the role of connecting Substations NEL and MUR after the load at MAN is transferred to the south metro system. In option 2, the loading on cable 2L40 is reduced, resulting in a lower failure probability. The sensitivity study indicates that a slightly lower failure probability of 2L40 would make option 2 less risky than the base case. Therefore, transferring the load at MAN to the south metro system while opening 2L46/53 was recommended as the operation mode for taking 2L39 out of service in the winter period since it was the lowest operation risk mode under this condition.

Table 7.3 EENS (MWh) for the base case and the two options to reduce loading on cable 2L40

	North system		South system		
	Oct.–Mar.	Apr.–Jun.	Oct.–Mar.	Apr.–Jun.	Total
Base case (No 2L39 but with the load transfer)	1939	887	1023	744	4593
Option 1 (T4 & T5 opened)	2724	1428	1023	744	5919
Option 2 (2L46/53 opened)	1734	1112	1023	744	4613
Option 2 with lower failure prob. of 2L40	1567	904	1023	744	4238

7.4.4   Selecting Operation Mode under the *N* – 2 Condition

Cable 2L40 was also in its end-of-life period, with a high failure probability so that it could fail during the replacement of 2L39. In other words, it was quite possible that both 2L39 and 2L40 could be out of service at the same time. Under this situation, the north metro system could no longer meet the single-contingency criterion. With 2L39 and 2L40 out of service (an *N* – 2 condition), one more single-component outage would essentially be a triple-outage event if the normal operation condition before the 2L39 replacement were taken as a reference. The system is not designed to meet the *N* – 3 criterion. However, the operators must be prepared for such a situation in operation planning. The purpose is not to find the operation mode satisfying the *N* – 3 criterion but, rather, to identify a relatively low-risk operation mode when both 2L39 and 2L40 are out of service.

Based on the power-flow studies, three representative options are selected and discussed as follows:

- Option 1: adding 2L40 out of service to the operation mode that has been recommended for 2L39 out of service (i.e., the load transfer at MAN to the south metro system, both 2L39 and 2L40 out of service, and 2L46/53 opened).
- Option 2: the same as option 1 except that 2L46/53 is closed. After both 2L39 and 2L40 are out of service, there may be too much power flowing through the north path if 2L46/53 is opened.
- Option 3: option 1 plus opening the overhead lines 60L2, 60L3, and 60L18 at Substation COK and the line 60L17 at Substation BND. Opening the four 60 kV circuits will greatly reduce the burden in the 230 kV system. This is because the three substations (CQM, ESC, and MRG) located at the right-most side in the diagram would be supplied only from the 60 kV system in the neighboring area. Also, the 60 kV substations along the line 60L17 would be supplied mainly from Substation COK through the 60 kV system. The effect of this measure is to reduce the operation risk in the downtown area when the condition with the two critical cables out of service becomes a basic operation state.

Table 7.4 shows the EENS indices for the three options. To be comparable, the EENS indices for the neighboring area must be included since option 3 is associated with the risk increase in the neighboring area. The EENS indices for the south metro system have been excluded from the comparison because they are the same for all the three options. Note that all the three options are viewed as normal states in the risk evaluation since the EENS indices are only caused by the subsequent random failures of other components in the system.

The following observations can be made from the results:

- Option 1 requires load curtailments even in the normal state and thus is not acceptable. In other words, opening 2L46/53 is a good measure for 2L39 out of service but no longer works for both 2L39 and 2L40 out of service.

Table 7.4 EENS (MWh) of the three options under the condition of both 2L39 and 2L40 being out of service

	North system		Neighboring system		
	Oct.–Mar.	Apr.–Jun.	Oct.–Mar.	Apr.–Jun.	Total
Option 1 (2L46/53 opened)	load curtailments required in the normal state for the winter peak period				
Option 2 (2L46/53 closed)	4778	2804	621	513	8716
Option 3 (2L46/53 opened plus four 60 kV lines opened)	2161	1012	1647	1425	6245

- Option 2 (2L46/53 closed) does not require any load curtailment in the normal state but generates very high operation risk, which is created by the random failures of other components. Many single-component failure events under the condition of both 2L39 and 2L40 out of service will lead to a large amount of load curtailments.

- Option 3 provides relatively low operation risk under the condition of both 2L39 and 2L40 being out of service. The essence of this option is a power supply transfer from the 230 kV system to the 60 kV system through the network configuration changes. The shift of system operation risk comes with the power supply transfer and the total risk has been reduced. This is due to the fact that the same loads will cause a different risk level if they are supplied through varied routes because the lines and transformers at different places have different capacity limits and failure probabilities.

- Opening the 60 kV circuits in option 3 increases the risk in the neighboring area. This is because the three substations at the right side in Figure 7.4 are now supplied only from the neighboring area whereas they originally were supplied from both the north metro and neighboring subsystems. This is thought of as a reasonable measure to deal with the high-risk situation in the downtown area, which is located within the north metro subsystem.

7.4.5 Summary

The example demonstrates an actual application of transmission risk evaluation in operation planning. The impacts of the load transfer and network reconfigurations/switching on the system risk were assessed and the lowest-risk operation modes under the different operation conditions were identified.

The study indicates that the load transfer at the MAN substation to the south metro area is an effective measure to reduce the operation risk in the north metro area when cable 2L39 is out of service for replacement. Opening 2L46/53 in the winter period (from October 2000 to March 2001) could provide the further improvement in the operation reliability. Under the condition of both 2L39 and 2L40 being out of service, the relatively low risk operation mode is to open 2L46/53 and the four 60 kV circuits that are connected to the neighboring area. This leads to a risk transfer from the north metro subsystem, which includes the downtown area, to the neighboring subsystem that is located in the suburbs, resulting in a decrease in the operation risk of the whole system. The system is not designed to satisfy the $N-2$ condition. The risk transfer from the most important area to the less important one is a rational measure in risk management under the $N-2$ condition.

7.5   EXAMPLE 2: A SIMPLE CASE BY HAND CALCULATIONS [76]

There are many simple cases in actual operation planning in which a risk evaluation can be quickly conducted through a spreadsheet or even hand calculations rather than using a complex computing program. The purpose of this example is to provide an insight into risk evaluation using the manual calculation method. The reader can also acquire useful ideas from the example about how to develop a small computing tool.

7.5.1   Basic Concept

The task of operation planning includes determination of a short-term operation scheme associated with only a few transmission components. In this case, it is possible to only consider a localized network structure and evaluate limited system states using the state enumeration technique. Such a risk evaluation process can be performed on a spreadsheet or through hand calculations.

The basic procedure includes the following steps:

- A localized network structure is separated from the full transmission system based on the nature of a problem.
- Consider two or more operation schemes for comparison. The schemes are often associated with different switching actions.
- Determine the system states to be enumerated for each scheme. In most cases, only single or double contingencies are considered.
- Evaluate the probabilities and consequences (load or energy curtailments) of the system states for each scheme. A failure event may occur at different points in time and time-dependent factors should be incorporated in the evaluation.

7.5.2   Case Description

This is the representative case that an operation planner often faces. As shown in Figure 7.5, the two bus loads are supplied through the three transmission cables in the normal operating state. The cables C1, C2, and C3 have capacity limits, respectively, of 400 MW, 170 MW, and 200 MW. The loads at busbars 1 and 2 are 240 MW and 120 MW. It can be seen that the system meets the single-contingency cri-

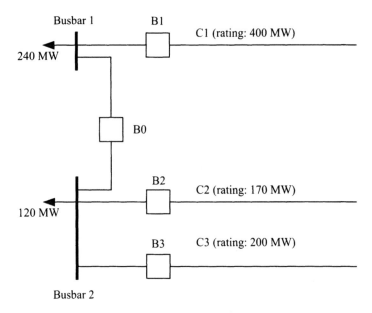

Figure 7.5 A localized network structure.

terion—a failure of any one cable will not lead to any overloading and no load curtailment is required.

Cable C3 is scheduled for maintenance for 2 days. During the maintenance, if cable C1 fails, there will be immediate and severe overloading on cable C2, causing damage. In order to avoid the possible damage, the following two operating schemes are considered:

Scheme 1: As shown in Figure 7.6, C2 is normally disconnected from busbar 2 by opening breaker B2 so that the two loads are supplied only through C1. When C1 fails, B2 is closed in order for C2 to pick up to 170 MW.

Scheme 2: As shown in Figure 7.7, breaker B0 is normally opened so that the load at busbar 1 is supplied through C1 and the load at busbar 2 through C2. When C2 fails, B0 is closed so that C1 can supply both the loads. When C1 fails, B0 is closed in order for C2 to pick up to 170 MW.

Both the schemes can avoid the possible damage on C2. The purpose of the risk evaluation is to identify which scheme is less risky to the power supply.

7.5.3 Study Conditions and Data

The study conditions are as follows:

- C3 is out of service for maintenance for 48 hours.
- There is no automatic switching scheme between C1 and C2 as the protection coordination is designed only for the normal operating state in which all three

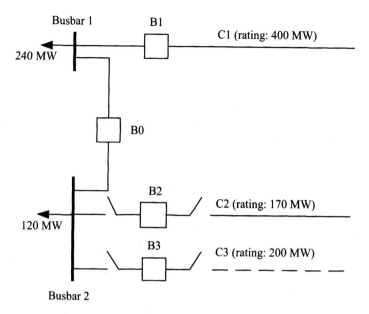

Figure 7.6   Operation Scheme 1 when C3 is out of service.

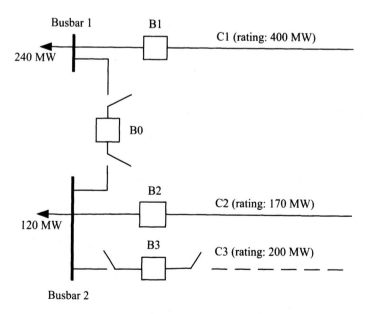

Figure 7.7   Operation Scheme 2 when C3 is out of service.

cables are in service. In other words, when C1 or C2 fails, a switching process associated with breaker B0 or B2 and their disconnect switches has to be carried out manually. It is estimated to take 1 to 2 hours to complete the switching process, including the signal transmission from the field to the control center, operator's judgment, and manual manipulation.

- Only the single outage of C1 or C2 is considered since the simultaneous failure of C1 and C2 makes the same contribution to the two schemes. In fact, the double-outage event within two days has an extremely small probability of occurrence and can be ignored in the example.
- Both the loads are industrial customers and have a flat load level in the period of 48 hours during the C3 maintenance.
- The C1 or C2 failure can take place at any time during the C3 maintenance. For the following hand calculations, one hour is used as an interval. A failure is assumed to occur at the beginning of each interval. If a small computer program is developed, or the calculation is conducted on a spreadsheet, the interval length can be reduced to increase calculation accuracy.

The data are given in Tables 7.5 and 7.6. The failure data of C1 and C2 are based on the historical statistics in the past 10 years.

7.5.4 Risk Evaluation

7.5.4.1 Calculating the Failure State Probability. The average failure frequency per hour is

$$f_{C1} = 1.2/8760 = 0.00013699 \text{ failures/hr}$$

$$f_{C2} = 2.3/8760 = 0.00026256 \text{ failures/hr}$$

The failure frequency per hour can be approximately used as the failure rate per hour since the repair times of C1 and C2 are very short compared to the mean time between failures, which is the reciprocal of the failure frequency. The repair times of C1 and C2 are longer than the return time of C3 (48 hours) so that a failure process will be terminated once the C3 maintenance is completed. The failure probability of C1 and C2 in one hour can be approximately estimated as

Table 7.5   Load and cable rating

Load at busbar 1	240 MW
Load at busbar 2	120 MW
Cable C1 rating	400 MW
Cable C2 rating	170 MW
Cable C3 rating	200 MW

Table 7.6   Failure and switching data

	Failure Frequency (failures/yr)	Repair time (h/failure)
Cable C1	1.2	51.2
Cable C2	2.3	62.3
B0 switching time		1 to 2 h
B2 switching time		1 to 2 h

$$P_{C1} = f_{C1} \times 1 = 0.00013699$$

$$P_{C2} = f_{C2} \times 1 = 0.00026256$$

There are the following two system failure states:

1. C1 failed and C2 not failed
2. C2 failed and C1 not failed

The probabilities of the two failure states are:

$$P_1 = 0.00013699 \times (1 - 0.00026256) = 0.00013695$$

$$P_2 = 0.00026256 \times (1 - 0.00013699) = 0.00026252$$

It should be noted that the simultaneous failure of C1 and C2 is excluded from the following calculations for both the schemes.

7.5.4.2 Evaluating EENS by Assuming One-Hour Switching Time

Scheme 1. Only the first system failure state needs to be evaluated for Scheme 1. There are two subevents associated with this failure state:

1. For the first hour following the C1 failure, both the loads have to be curtailed. The expected energy not supplied (EENS) is calculated as

$$EENS_a = (240 + 120) \times 0.00013695 \times 1 = 0.049302 \text{ MWh}$$

2. In the subsequent period, B2 is closed and C2 picks the loads up to 170 MW. The C1 failure can happen at any point of time within the 48 hours. If it takes place at the beginning of the first hour, the length of the subsequent period is 47 hours; if it takes place at the beginning of the second hour, the length is 46 hours; and so on. If it takes place at the beginning of the 48th hour, the length is zero hours. The EENS is calculated as

$$EENS_b = (240 + 120 - 170) \times 0.00013695 \times (47 + 46 + \ldots + 2 + 1 + 0)/48$$
$$= 0.611482 \text{ MWh}$$

The total EENS for Scheme 1 is

$$\text{EENS}(1) = EENS_a + EENS_b = 0.049302 + 0.611482 = 0.660784 \text{ MWh}$$

Scheme 2. Both the system failure states need to be evaluated for Scheme 2. Under the first failure state, there are two subevents:

1. For the first hour following the C1 failure, the load at busbar 1 is curtailed. The EENS is calculated as

$$EENS_a = 240 \times 0.00013695 \times 1 = 0.032868 \text{ MWh}$$

2. In the subsequent period, B0 is closed and C2 picks the loads up to 170 MW. The calculation of the EENS is the same as that in the subevent 2 for Scheme 1:

$$EENS_b = (240 + 120 - 170) \times 0.00013695 \times (47 + 46 + \ldots + 2 + 1 + 0)/48$$
$$= 0.611482 \text{ MWh}$$

Under the second failure state, there are also two subevents:

1. For the first hour following the C2 failure, the load at busbar 2 is curtailed. The EENS is calculated as

$$EENS_c = 120 \times 0.00026252 \times 1 = 0.031502 \text{ MWh}$$

2. In the subsequent period, B0 is closed and C1 picks all the loads up so that no curtailment is required:

$$EENS_d = 0$$

The total EENS for Scheme 2 is

$$EENS(2) = EENS_a + EENS_b + EENS_c + EENS_d$$
$$= 0.032868 + 0.611482 + 0.031502 + 0 = 0.675852 \text{ MWh}$$

7.5.4.3 *Evaluating EENS by Assuming Two-Hour Switching Time.* The calculation procedure is similar to the case of assuming one-hour switching time except for the time length of the subevents.

Scheme 1. The system failure state to be considered is the same—C1 failed but C2 not failed. The EENS indices for the two subevents before and after the switching action are calculated as follows.

1. Before B2 is closed, both the loads have to be curtailed. The time length of the load curtailments depends on when the failure takes place. If the failure occurs before the beginning of the 47th hour, the time length is two hours. If it occurs at the beginning of the 48th hour, the length is only one hour since C3 is assumed to return from maintenance by the end of the 48th hour:

$$EENS_a = (240 + 120) \times 0.00013695 \times (2 \times 47/48 + 1 \times 1/48) = 0.097577 \text{ MWh}$$

2. After B2 is closed, C2 picks the loads up to 170 MW. If the C1 failure takes place at the beginning of the first hour, the length of the subsequent period is 46 hours; if it takes place at the beginning of the second hour, the length is 45 hours; and so on. If it takes place at the beginning of the 47th or 48th hour, the length is zero hours:

$$EENS_b = (240 + 120 - 170) \times 0.00013695 \times (46 + \ldots + 2 + 1 + 0 + 0)/48$$
$$= 0.586003 \text{ MWh}$$

The total EENS for Scheme 1 is

$$EENS(1) = EENS_a + EENS_b = 0.097577 + 0.586003 = 0.683580 \text{ MWh}$$

Scheme 2. The two system failure states to be evaluated are the same—C1 failed and C2 not failed, and C2 failed and C1 not failed. Under the first failure state, the EENS indices for the two subevents are calculated as follows:

1. Before B0 is closed, the load at busbar 1 is curtailed:

$$EENS_a = 240 \times 0.00013695 \times (2 \times 47/48 + 1 \times 1/48) = 0.065051 \text{ MWh}$$

2. After B0 is closed, C2 picks the loads up to 170 MW. The EENS is the same as that in the subevent 2 of Scheme 1:

$$EENS_b = (240 + 120 - 170) \times 0.00013695 \times (46 + \ldots + 2 + 1 + 0 + 0)/48$$
$$= 0.586003 \text{ MWh}$$

Under the second failure state, the EENS indices for the two subevents are calculated as follows:

1. Before B0 is closed, the load at busbar 2 is curtailed:

$$EENS_c = 120 \times 0.00026252 \times (2 \times 47/48 + 1 \times 1/48) = 0.062349 \text{ MWh}$$

2. After B0 is closed, C1 picks all the loads up so that no curtailment is required:

$$EENS_d = 0$$

The total EENS for Scheme 2 is

$$EENS(2) = EENS_a + EENS_b + EENS_c + EENS_d$$
$$= 0.065051 + 0.586003 + 0.062349 + 0 = 0.713403 \text{ MWh}$$

7.5.5 Summary

The results in Section 7.5.4 indicate that Schemes 1 and 2 have almost the same operation risk if the switching time of B0 and B2 is one hour. Scheme 2 becomes slightly more risky if the switching time is longer. The total EENS values are very small, indicating that the system operation risk during the two days of the cable C3 maintenance is low regardless of which scheme is used.

The example demonstrates how the risk evaluation for a simple case in transmission operation planning can be performed using hand calculations. The procedure is easy to carry out on a spreadsheet. Risk assessment does not always have to be complex. If a sophisticated computer program is not available, operation planning engineers can still conduct similar risk studies. The key is determination of the system states to be evaluated and the subevents related to each state that are generally divided according to switching actions. The evaluation is associated with the three quantities of probability, consequence (load curtailments), and time length. It is important to appreciate that the probability of a system failure state does not only depend on failure components. It is the product of the failure probabilities of failure components and the success probabilities of nonfailed components. The load curtailments due to a subevent should be able to be obtained through simple calculations or even direct observation. This is a necessary condition in order to use the hand calculation method. The time length of a subevent relies on the moment at which a switching action takes place, the duration needed by the switching action, and the total time frame considered in the study.

7.6 CONCLUSIONS

This chapter discusses the risk evaluation method and its applications in transmission system operation planning. There are fundamental differences between the risk assessments in system long-term development planning and operation planning for a short time period. The latter has to focus on the simulation of the operation measures. These include the load transfers, generation pattern changes, temporary network reconfigurations, and switching actions. The network configuration changes for a relatively long time length and the switching actions based on individual system states are often related to the noncoherence phenomenon in transmission system risk assessment. A method or tool that assumes coherence generally cannot be used in transmission risk assessment, particularly in the risk evaluation of transmission operation.

The risk evaluation of operation modes is usually associated with a partial-risk transfer from one area to another or from one voltage level to another. These are not the same as transfers with equal amounts of risk. The total risk level of the whole system may be reduced in the transfer. The rationality and tolerance level of risk transfer to an area can be assessed through a quantified risk index.

An operation engineer has to make decisions on a daily, weekly, or monthly basis. There are many simple cases in which an operation scheme is only related to a localized network structure and the consequences of failure events can be directly observed. In these cases, the risk evaluation can be still quickly performed using a hand or spreadsheet calculation method if an appropriate computer program is not available.

CHAPTER 8

APPLICATION OF RISK EVALUATION TO GENERATION SOURCE PLANNING

8.1 INTRODUCTION

Generation source planning is a classic and complex task in utilities. The basic purpose is to determine the enhancement of generation sources to meet load growth with acceptable adequacy. Generally, the planning process does not consider transmission networks but focuses on the balance between generation and load. The traditional generation risk evaluation is performed under this assumption. Considerable discussion on the subject can be found in many publications [1, 3, 19, 20, 38, 41]. The intent of this chapter is not to redescribe the materials that have been addressed elsewhere. Instead, a reliability planning method for generation sources under the constraints of transmission networks is presented. With the introduction of the transmission network, the impact of a generation source on system risk depends on not only its size but also its location in the system. The constraint due to the transmission network is particularly important in planning the local generation sources like independent power producers (IPP) or cogenerators. The basic concept is illustrated in Section 8.2 and the simulation approach in Section 8.3. The method presented can be applied in the following areas:

- The comparison between different generation source schemes
- The comparison between generation enhancements and transmission expansions (integrated generation and transmission planning)
- The IPP (independent power producer) or cogeneration planning
- The study of the impact on system risk of any decision associated with generation equipment (maintenance, refurbishment, retirement, etc.)

Risk Assessment of Power Systems. By Wenyuan Li
ISBN 0-471-63168-X © 2005 the Institute of Electrical and Electronics Engineers, Inc.

After going through Sections 8.2 and 8.3, the reader will find that the applications of the first two aspects are relatively straightforward. Two examples are given in Sections 8.4 and 8.5. The first one is an application for determining the size and location of cogenerators and the second one is a study of the retirement decision of a generator in a regional system.

8.2 PROCEDURE FOR RELIABILITY PLANNING [33, 53, 59, 61]

The reliability planning of generation sources includes the following aspects:

1. Select the feasible generation source schemes based on societal, environmental, and regulatory considerations. If integrated resource planning is performed, the feasible transmission reinforcement schemes are also selected.
2. Determine a short list of planning alternatives by means of the technical analysis and risk sensitivity evaluation. The technical analysis includes the system power flow, fault level, transient stability, and technical feasibility. The risk sensitivity evaluation is conducted to examine the impacts on system risk of the generation additions with varied sizes and locations. This can be performed using a composite generation and transmission system risk assessment method. The short list covers the size ranges and locations of the potential generation additions.
3. Conduct stochastic simulations of generation costs and risk costs for the alternatives in the short list using a Monte Carlo based approach. This will be discussed in Section 8.3.
4. Perform the overall economic analysis including the investment, operation, and risk costs. The analysis requires the calculations of the cash flows and present values of the three costs over a planning time span such as 10 to 20 years. The basic concepts of the cash flow and present value have been discussed in Sections 6.2.2 and 6.2.3. The annual equivalent investment can be calculated using Equations (6.2) and (6.3), and the present value using Equation (6.4). The operation cost includes the generation (fuel), maintenance, and administration costs. The generation and risk costs can be obtained through the production simulation in Step 3.
5. There are two criteria used to determine the best alternative. If a utility invests in the added generators and operates the whole system, including both generation and transmission, the best alternative is the one that requires the lowest total cost:

$$\min T = I + O + R \tag{8.1}$$

where I, O, and R are the investment, operation, and risk costs, respectively.

Under the deregulation environment, however, the generation and transmission may belong to different owners. Also, nonutility generations (cogenerators or independent power producers) are allowed to have an access to a utility system. This brings both benefits and costs to the utility, although the investment and operation costs of the nonutility generation are not paid by the utility. In this case, the criterion to select the best alternative to the utility should be maximization of the net benefit:

$$\max NB = B - C \qquad (8.2)$$

where NB is the net benefit. B and C are, respectively, the benefit and cost to the utility due to a generation addition. B and C have different compositions depending on the case under study. If multiple years are considered, this formula can use either present value or direct sum of yearly benefits and costs. The example of cogeneration planning given in Section 8.4 will provide more particulars.

8.3 SIMULATION OF GENERATION AND RISK COSTS

The Monte Carlo sampling technique combined with a minimum cost assessment model is used to conduct the simulation of generation and risk costs.

8.3.1 Simulation Approach

The simulation approach includes the following basic steps:

1. The multiple step annual load model is created, which eliminates the chronology and aggregates load states using hourly load records.
2. The system states at a load level are selected using a Monte Carlo simulation process. The system under consideration is a composite generation and transmission system. Generating units are generally modeled using multiple state random variables, including the up, down, and derated states, whereas transmission components are represented using two-state variables (only up and down states) or are assumed to be 100% reliable, depending on the case. The inclusion of a transmission network is necessary since the effect of a generation addition may be limited due to constraints of the transmission network.
3. The minimization cost model given in the next section is solved to obtain the generation allocations, generation costs, load curtailments, and interruption costs at buses.
4. Steps 2 and 3 are repeated until the convergence for each load level.
5. The results for all the load levels are weighted by their probabilities to calculate the annual indices of the expected generation and risk costs.

The simulation approach is shown in Figure 8.1.

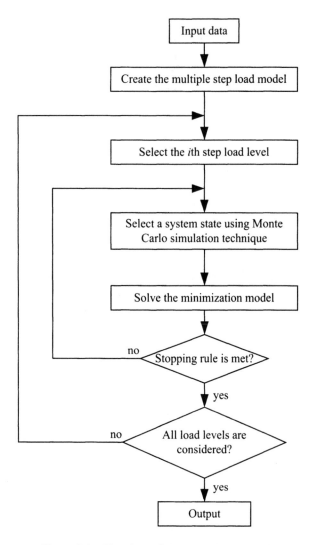

Figure 8.1 Flowchart of the simulation approach.

8.3.2 Minimization Cost Model

For a sampled system state, the following minimization model is used to calculate the output allocations among generators, possible load curtailments at different load buses, generation costs at each generator bus, and interruption costs at each load bus:

$$\min \sum_{i \in NG} B_i PG_i + \sum_{i \in ND} W_i C_i \tag{8.3}$$

subject to:

$$T_n = \sum_{i=1}^{NS} A_{ni}(PG_i - PD_i + C_i) \qquad (n = 1, \ldots, L) \qquad (8.4)$$

$$\sum_{i \in NG} PG_i + \sum_{i \in ND} C_i = \sum_{i \in ND} PD_i \qquad (8.5)$$

$$PG_i^{\min} \le PG_i \le PG_i^{\max} \qquad (i \in NG) \qquad (8.6)$$

$$0 \le C_i \le PD_i \qquad (i \in ND) \qquad (8.7)$$

$$|T_n| \le T_n^{\max} \qquad (n = 1, \ldots, L) \qquad (8.8)$$

where PG_i is the generation variable at the ith bus; PD_i is the load at the ith bus; C_i is the load curtailment variable at the ith bus; T_n is the line flow on the nth branch; A_{ni} is the element of the relation matrix between branch flows and power injections for the sampled system state; $PG_i^{\max}$ and $PG_i^{\min}$ are the upper and lower limits of generation variables; $T_n^{\max}$ is the rating of the nth branch; W_i is the weighting factor reflecting load bus importance; B_i is the unit generation cost for the ith generator bus; ND is the set of load buses; NG is the set of generation buses; NS and L are the numbers of buses and branches in the system, respectively.

It can be seen that the objective of the model is to minimize the sum of generation costs and interruption costs while satisfying the power balance, linearized load flow relationships, branch ratings, and generating unit limits. The smallest W_i is selected to be larger than the largest B_i so that the model can automatically redispatch the generations before any load curtailment in minimizing the total cost. The load curtailments only occur when they are unavoidable in order to meet all the constraints.

It is apparent that not only the unit generation cost B_i of the generators in the objective function but also their locations reflected in the constraints have impacts on the total generation and risk costs.

8.3.3 Expected Generation and Risk Costs

The expected annual interruption cost (EAIC) and expected annual generation cost (EAGC) are calculated using Equation (8.9)–(8.12).

$$EAIC_i = \sum_{j=1}^{NL} \sum_{k \in S_j} \alpha_i \cdot C_{ik} P_k \cdot T_j \qquad (8.9)$$

$$EAIC = \sum_{i \in ND} EAIC_i \qquad (8.10)$$

$$EAGC_i = \sum_{j=1}^{NL} \sum_{k \in S_j} B_i \cdot G_{ik} P_k \cdot T_j \tag{8.11}$$

$$EAGC = \sum_{i \in NG} EAGC_i \tag{8.12}$$

where $EAIC_i$ and $EAIC$ are the expected annual interruption costs for the ith load bus and for the system; $EAGC_i$ and $EAGC$ are the expected annual generation costs for the ith generator bus and for the system; P_k is the probability of the kth system state, which is determined in the Monte Carlo simulation; C_{ik} is the load curtailment (MW) at the ith load bus in the kth system state; α_i is the unit interruption cost ($/MWh) at the ith load bus; G_{ik} is the generation output (MW) at the ith generator bus in the kth system state; B_i is the unit generation cost ($/MWh) at the ith generator bus; S_j is the set of sampled system states at the jth load level in the multiple-step load model; T_j is the time length of the jth load level; and NL is the number of load levels.

Note that α_i may or may not be the same as W_i in Equation (8.3). In many cases, utilities may not be able to differentiate the unit interruption costs at different buses and the identical α_i can be used for all buses, whereas W_i at each bus only needs to be specified in terms of its relative magnitude to reflect the importance of the bus.

8.4 EXAMPLE 1: SELECTING LOCATION AND SIZE OF COGENERATORS [33]

It is straightforward to utilize the concept and the method presented in Sections 8.2 and 8.3 for utility-owned generators. However, additional considerations are needed in the application for the access of nonutility generators to a utility system. This example shows the latter case.

8.4.1 Basic Concept

In planning the access of nonutility generators, a utility has to go through a negotiation process with the candidate cogenerators. This requires a full understanding of cogeneration effects on the utility system in the economic, environmental, and technical areas. The access of nonutility generators to the utility system has both negative and positive impacts on the utility. The negative aspects include the loss of base loads and associated revenue, risk to ratepayers from reliance on the cogenerated power, and additional operating problems such as the increased fault current, voltage flicker, control and communication issues, and possible system instability. The benefits to the utility include the improvement in regional system reliability, possible reduction in network losses, and savings in generating capacity costs. From a societal perspective, the cogenerators may also provide regional economic and environmental benefits.

After feasibility investigations and environmental, technical, and regulatory assessments, several potential candidates can be short-listed. The basic task of the utility in the cogeneration planning is to prioritize the candidates from the viewpoint of economic and reliable operation and identify the most appropriate location and size of cogenerators.

The impact of a cogeneration facility on the utility system reliability is an important factor in determining its location and size. If the power from main generation sources can effectively reach a local area in any contingency state, a cogenerator in the area may only provide limited improvement in area system reliability. Otherwise, if the outside power sources cannot effectively support the area due to transmission network constraints in the contingency states, the local cogenerator will be able to offer a high benefit to the area system reliability. The impacts of cogeneration should, therefore, be evaluated from the composite generation and transmission system, not just from the balance between generating capacities and load demands.

The purpose of the overall economic analysis is to maximize the net benefit to the utility. The reductions in the network losses, system risk, and demand of total generation capacity are the three main benefits. The major negative impact of a cogeneration facility to the utility is due to the buyback energy since it is generally associated with the loss of revenue. In this application, Equation (8.2) can be rewritten as:

$$\max NB = RR + LR + GR - EC \qquad (8.13)$$

where NB represents the net benefit, RR is the reduction in the system risk cost, LR the reduction in the network loss cost, GR the reduction in the requirement of total generation capacity cost, and EC the equivalent cost to the utility, which is the loss of revenue due to the access of cogeneration.

It should be noted that cogeneration in a local area basically does not change the average generation production cost of the utility. The impact of cogeneration access on the reduction in generation production of the utility is considered through the two components of GR and EC in Equation (8.13). In this application, therefore, the minimization model in Section 8.3.2 is only used to calculate the expected system risk cost. In order to obtain the reduction in the system risk cost, the simulation process has to be conducted twice for each cogeneration scheme, one without and another with the cogeneration. The network losses are calculated using a power-flow-based computing tool that can handle a load curve. Similarly, the loss evaluation should also be performed twice, with and without the cogeneration, to obtain the loss reduction.

8.4.2   System and Cogeneration Candidates

The following is an actual application of probabilistic cogeneration planning at a utility. The single-line diagram of a regional system in the study is shown in Figure 8.2. The region is a load center far away from major generation sources. In order to incorporate the network constraint between the region and the generation sources, the system evaluated includes the full local 230 kV and 69 kV subsystems and the 500 kV transmission network connecting the generation sources to the region. To

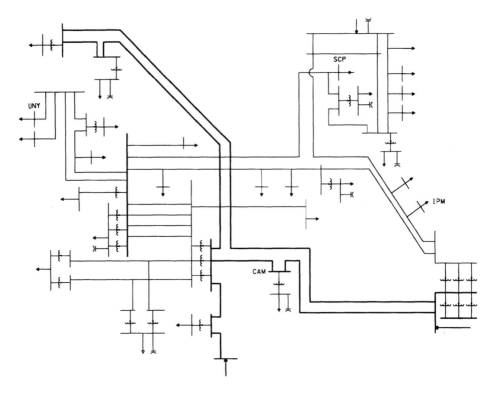

Figure 8.2 Regional system and locations of potential cogenerators.

highlight the system risk in the region, only the component failures in the region are simulated and all the components outside the region have been assumed to be 100% reliable. As a result, all load curtailments are located inside the region. The failure data of the transmission components are retrieved from the database of historical records, whereas the data for the cogenerators are based on an annual report of the Canadian Electricity Association (CEA) [80]. The initial load level corresponds to the 1995 load profile and it has been assumed that industrial loads remain flat but nonindustrial loads have 2% of load growth for each following year. The annual load-duration curves for different years are assumed to have the same shape. This means that 8760 hourly loads in a different year increase in the same proportion as the peak load in that year. The coefficient of variance for the EENS (expected energy not supplied) index was used as the convergence criterion in the Monte Carlo simulation and was set to be 0.05 in all the studied cases.

Based on the initial feasibility investigation, the potential locations and capacity ranges of cogenerators in the region are as follows:

1. A 40–50 MW hog-fueled cogeneration facility of industrial customer A is connected to the 69 kV system at Substation SCP.

2. An 80 or 150 MW gas-fueled cogeneration facility of industrial customer B is connected to the 69 kV system at Substation IPM.

3. The same as 2 except that the cogenerator is connected to the 230 kV system through a line to Substation CAM.

4. A 50–80 MW gas-fueled cogeneration facility of industrial customer C is connected to the 69 kV system at Substation UNY.

8.4.3   Risk Sensitivity Analysis

The risk sensitivity analysis was conducted using an incremental increase of cogeneration from 10 MW up to 150 MW at each of the four possible locations. The system risk of each case for the two load levels in 1995 and 2003 was evaluated. In this step, the annualized indices were calculated at the peak load and expressed on a one-year basis. The annualized indices are acceptable for the purpose of relative comparisons. The annual indices considering a yearly load curve will be used in the following economic analysis. The annualized expected energy not supplied (EENS) indices for the four possible cogeneration locations are shown in Tables 8.1 and 8.2. Figures 8.3 and 8.4 are their graphical expressions.

It can be seen from the results that the cogenerator connected to Substation SCP provides very limited improvement in the system risk. The cogenerator of customer B connected to the 69 kV system at Substation IPM provides a much

Table 8.1   Annualized EENS (MWh/yr) for the four cogeneration locations at 1995 load level

Cogenerator Size (MW)	Substation to be connected			
	SCP	IPM	CAM	UNY
0	1224	1224	1224	1224
10	1178	1197	1143	1070
20	1142	1175	1067	939
30	1121	1154	991	885
40	1110	1134	914	830
50	1090	1117	838	828
60	1089	1108	762	828
70	1079	1099	691	828
80	1068	1090	621	828
90	1057	1080	555	828
100	1047	1072	554	828
110	1036	1063	554	828
120	1026	1054	553	828
130	1015	1045	553	828
140	1005	1036	553	828
150	1005	1032	553	828

Table 8.2 Annualized EENS (MWh/yr) for the four cogeneration locations at 2003 load level

Cogenerator Size (MW)	Substation to be connected			
	SCP	IPM	CAM	UNY
0	9400	9400	9400	9400
10	8829	9047	7780	8157
20	8471	8725	6524	6978
30	8174	8478	5472	6158
40	7917	8260	4420	5494
50	7674	8050	3371	5270
60	7444	7845	2343	5270
70	7232	7650	2183	5270
80	7006	7475	2068	5270
90	6808	7308	1957	5270
100	6597	7141	1846	5270
110	6422	6974	1754	5270
120	6273	6838	1666	5270
130	6124	6812	1579	5270
140	5974	6812	1492	5270
150	5957	6812	1406	5270

smaller improvement in the system risk compared to its connection to the 230 kV system at Substation CAM. The 50 MW cogenerator connected to the 69 kV system at Substation UNY provides a relatively significant improvement in the system risk but a larger unit (beyond 50 MW) at the same substation does not lead to further improvement. A similar situation can be observed for the cogenerator connected to the 230 kV system at Substation CAM; that is, the addition

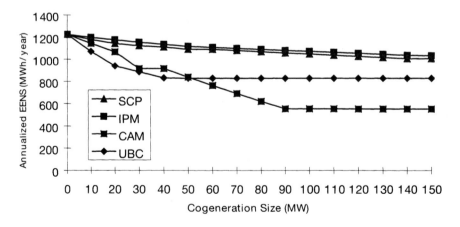

Figure 8.3 Annualized EENS for four cogeneration locations at 1995 load level.

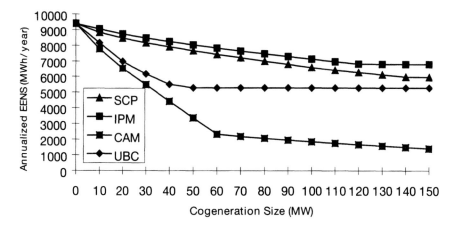

Figure 8.4   Annualized EENS for four cogeneration locations at 2003 load level.

of the first 80 MW cogenerator provides a relatively large improvement in the system risk, whereas the addition beyond 80 MW has a very limited further impact.

The observations can be explained as follows. Addition of a cogenerator can lighten or even eliminate some system problems caused by the outages that occur near the cogeneration location. The cogeneration cannot alleviate the system problems due to outages far away because of the transmission line constraint. This analysis suggests that the system risk is sensitive to locations and sizes of cogenerators. In other words, an unsuitably located cogenerator may not provide any effective improvement in the system risk, although it increases the total generation capacity in the system.

8.4.4 Maximum Benefit Analysis

The results obtained by the risk sensitivity analysis do not support the cogenerators connected to the 69 kV system at Substations SCP and IPM. The network loss calculations, on the other hand, indicate that adding the same amount of cogeneration at the four possible locations provides basically the same loss reduction in the region. Although cogeneration beyond 80 MW connected to the 230 kV system at Substation CAM provides almost no further improvement to the system reliability compared to the first 80 MW, it would reduce more network losses. Therefore, the following two cogeneration schemes were selected for further assessment:

1. 50 MW connected to the 69 kV system at Substation UNY and 80 MW connected to the 230 kV system at Substation CAM
2. 50 MW connected to the 69 kV system at Substation UNY and 150 MW connected to the 230 kV system at Substation CAM

Table 8.3 Annual EENS (MWh/yr) for the base case and two cogeneration schemes

Scheme	Load Level Year				
	1995	1997	1999	2001	2003
Base case	315	351	398	471	589
Scheme 1	205	216	228	245	267
Scheme 2	204	212	222	234	249

The annual EENS indices considering the yearly load curve for the base case without cogeneration and the two cogeneration schemes for the 1995 to 2003 load levels are presented in Table 8.3.

In the economic analysis, the period of eight years from 1995 to 2002 was considered. An average unit interruption cost of $6.3/KWh was estimated from the customer damage functions and customer-sector compositions in the region. The base system case is used as a reference. The reductions in the annual EENS index and system risk cost due to the two cogeneration schemes are shown in Tables 8.4 and 8.5, respectively. Note that the present value method is not used in this example, that is, the total cost is the direct sum of the annual costs of the eight years.

The load-flow-based method considering the load curve was used to calculate the network losses in the base case and two cogeneration schemes, and then the reductions in the network loss cost due to adding the cogenerators were assessed. The loss cost reduction is generally associated with capacity requirement reduction and energy loss reduction. The results are shown in Tables 8.6 and 8.7. It is noteworthy that the reductions in the network loss and loss cost do not monotonically change with the years because the load forecast does not have a monotonic pattern. The same network loss reduction does not necessarily lead to the same loss cost reduction in a different year since the rate associated with the energy cost varies with the years due to a few factors, including the inflation rate.

The access of a cogenerator to the utility system can reduce the system capacity requirements in generation–demand balance planning. This reduction, however, is

Table 8.4 Reduction in the annual EENS index (MWh/yr) due to the two cogeneration schemes

Year	Scheme 1	Scheme 2
1995	110	111
1996	122	125
1997	135	139
1998	153	158
1999	170	176
2000	198	207
2001	226	237
2002	274	288

Table 8.5 Reduction in the system risk cost (k$/yr) due to the two cogeneration schemes

Year	Scheme 1	Scheme 2
1995	693	699
1996	769	788
1997	851	876
1998	964	995
1999 .	1071	1109
2000	1247	1304
2001	1424	1493
2002	1726	1814
Total	8745	9078

Table 8.6 Reduction in the average network loss (MW) due to the two cogeneration schemes

Year	Scheme 1	Scheme 2
1995	5.8	8.1
1996	5.8	8.6
1997	6.5	9.7
1998	7.0	9.9
1999	7.8	11.3
2000	6.4	9.1
2001	6.2	9.0
2002	5.3	7.8

Table 8.7 Reduction in the network loss cost (k$/yr) due to the two cogeneration schemes

Year	Scheme 1	Scheme 2
1995	571	797
1996	589	874
1997	682	1018
1998	760	1074
1999	876	1269
2000	744	1057
2001	745	1083
2002	660	972
Total	5627	8144

not the megawatt value that can be provided by the cogenerator. The same capacity at major generation sources and cogeneration facilities does not create the same effect on system operation and system risk. The effect due to the cogenerator is limited to the local area, whereas that due to the major generation sources is more widespread and global. Generally, a discount coefficient is introduced to reflect this difference. The coefficient of 0.8 was used in this application.

In the case of Scheme 1, 50 MW of cogeneration is added at Substation UNY and 80 MW at Substation CAM. The system generation capacity requirement is reduced by $(50 + 80 - 52) \times 0.8 = 62.4$ MW, where 52 MW is the load of the customer providing the cogeneration. The 52 MW is deducted from the system capacity reduction because the utility makes a commitment to guarantee the power supply for the customer when its cogeneration facility fails. Based on the rate of the generation capacity ($34k/MW/year) defined in the capital planning guideline of the utility, the reduction in the capacity cost is $62.4 \times 34 = 2122$ k$/year. In the case of Scheme 2, 50 MW of cogeneration is added at Substation UNY and 150 MW at Substation CAM. The reduced system capacity requirement is $(50 + 150 - 52) \times 0.8 = 118.4$ MW and the corresponding capacity cost reduction is $118.4 \times 34 = 4026$ k$/year.

Economically, as mentioned earlier, the major negative impact of cogeneration to the utility is due to the energy bought back. In this application, the utility has a policy of buying back only 25% of nonutility generation outputs. The remaining 75% of the outputs are accessible to the system but have to be used for load displacement (wheeling to other customers). The energy sold to the utility by the cogenerator owners is essentially resold to other customers. The utility makes no profit from the resale. If the utility produces the same amount of energy and sells it to customers, however, it has a profit of $32.75/MWh. This means that the access of the cogenerators to the system results in a net revenue reduction to the utility. According to the historical statistics, a gas-fueled generator of 50 MW and over has an average unavailability of 0.35 including forced, planned, and maintenance outages. With the data above, in the case of Scheme 1, the net revenue reduction is $(1 - 0.35) \times 8760 \times (50 + 80 - 52) \times 32.75 \times 0.25 = 3636$ k$/year. In the case of Scheme 2, the net revenue reduction is $(1 - 0.35) \times 8760 \times (50 + 150 - 52) \times 32.75 \times 0.25 = 6900$ k$/year.

By applying Equation (8.13), the net benefits to the utility for the two cogeneration schemes in the eight years from 1995 to 2002 are calculated as follows:

Scheme 1:

$$\text{Net benefit} = 8745 + 5627 + (2122 - 3636) \times 8 = 2260 \text{ k\$}$$

Scheme 2:

$$\text{Net benefit} = 9078 + 8144 + (4026 - 6900) \times 8 = -5770 \text{ k\$}$$

It is interesting to note that Scheme 1 provides the utility with a positive benefit, whereas Scheme 2 provides a negative one. The most appropriate locations and sizes of cogenerators for the utility are therefore 50 MW connected to the 69 kV system at Substation UNY and 80 MW connected to the 230 kV system at Substation CAM.

8.4.5 Summary

This example demonstrates the application of risk evaluation in generation source planning. Besides the basic method, the additional analyses are required in planning the access of nonutility generators to the utility system. Generally, a nonutility generator creates both positive and negative impacts on the utility system. The impacts vary for different locations and sizes. In the given example, the two locations of cogenerator lead to very limited improvement in the system risk. At the other two locations, the cogenerator can decrease the system risk, but when the size is beyond 50 MW or 80 MW, it no longer provides further improvement. In the two cogeneration schemes selected from the preliminary risk sensitivity analysis, one can provide the utility with the positive benefit, whereas the other one results in the utility's economic loss.

It should be appreciated that the access of cogenerators to the utility system is also associated with other aspects such as operation issues and wheeling agreements. In other words, reliability planning is only one portion of the whole problem. However, the risk sensitivity analysis and the overall economic assessment, including the risk cost, provide important information for decision making. If a zero or negative-benefit cogeneration scheme (such as Scheme 2) is presented in the negotiation, for example, a high access charge fee should be considered. Alternately, the proportion of energy bought back should be decreased.

8.5 EXAMPLE 2: MAKING A DECISION TO RETIRE A LOCAL GENERATION PLANT [68]

The task of generation planning is not limited to the selection of new generation sources. Retirement of an aged generating unit is often a difficult issue in the decision-making process. The example below illustrates an application of risk evaluation in this regard.

8.5.1 Case Description

The study was conducted in 1999. The case was associated with a decision about the retirement of a generator that was located in the north region of an island. This was a relatively unimportant region and all loads were supplied through the 60 kV radial transmission lines as shown in Figure 8.5. The total load in the region was 54 MW and the power source was mainly from the GLD substation that was connected to the grid system. The local generating plant with a 50 MW generator was used as a standby power source. The generator had been operated for many years and reached its end-of-life stage. One of the two turbines had failed, resulting in reduc-

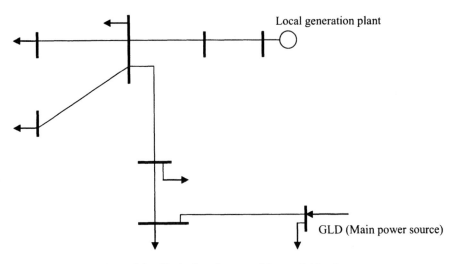

Figure 8.5   Single-line diagram of the north island area.

tion of its actual capacity to 25 MW. The unit had relatively many records of startup failures. The question faced by the utility was: should the generator be fully retired or upgraded back to 50 MW for continuous operation?

8.5.2   Risk Evaluation

The study conditions were as follows:

- The 1999 load profile was used. According to the load forecast, there was no load growth in the region.
- The annual load curve was based on the hourly load records of the local customers in the region.
- The failure data of the overhead lines were obtained from the database, which were the average statistics in the 8 years from 1990 to 1997.
- The utilization forced outage probability (UFOP) of the local generator was used rather than the traditional forced outage rate (FOR). The UFOP includes the effects of both forced outages and startup failures.

The following three alternatives were considered in the study:

1. The local generating unit is retired. There is no longer standby generation available and all the loads in this region would be supplied from the GLD substation.
2. The local generating unit is fixed so that it could be operated at the 50 MW capacity level for 10 more years. In other word, its retirement would be de-

layed until 2009. According to the historical statistics from the same type of units, the average UFOP for a 50 MW gas turbine generator is 12.2% and the average repair time is 56.04 hours. These are used as the normal outage data for the fixed generating unit.

3. The local generating unit maintains its status as is (the status in 1999), that is, at the 25 MW capacity with a higher UFOP. This means that no upgrade is carried out and only regular maintenance is performed so that the situation of the local generating unit would not become worse. The outage data for this status of the unit are UFOP = 23.38% (almost double the normal value) and the average repair time = 181.91 hours (about three times the normal value).

Table 8.8 shows the system risk indices for the three alternatives. The EENS indices (expected energy not supplied) for the three alternatives are 96 MWh/year, 30 MWh/year, and 65 MWh/year, respectively. The implications of the EFLC (expected frequency of load curtailment) and EDLC (expected duration of load curtailment) indices are explained as follows. For the alternative of retiring the local generating unit, the north island system would, on average, experience 0.54 occurrences of loss of load per year and the total duration of loss of load would be 3.03 hours per year. For the alternative of fixing the local generating unit, the events of loss of load would be reduced to 0.22 occurrences per year and the total duration of loss of load to 1.47 hours per year. For the alternative of keeping the local generating unit as is, the indices are 0.48 occurrences/year and 2.42 hours/year, respectively.

The following observations are made:

- The risk indices for the unit retirement option are still small and acceptable, although upgrading the unit can provide further improvement.
- If the local generating unit is retired, it would not cause significant deterioration in the system reliability compared to the option of keeping its status as is. This is mainly due to the fact that the overhead lines in the region have high availability according to the historical outage records, although they are in a radial structure.

8.5.3 Total Cost Analysis

The present value method of the total cost is applied. The time frame is the 11 years from 1999 to 2009.

Table 8.8 System risk indices for the three alternatives

Alternative	EENS (MWh/yr)	EFLC (occurrences/yr)	EDLC (h/yr)
Unit retired	96	0.54	3.03
Unit fixed (50 MW)	30	0.22	1.47
Unit kept as is (25 MW)	65	0.48	2.42

8.5.3.1 Investment Cost. Capital costs are not entailed with the two alternatives of retiring the unit and keeping it as is, whereas the alternative of fixing the unit requires $1200k in total, with $300k in 1999 and $900k in 2000.

Decommissioning costs for the alternative of retiring the unit are $450k in total, with $100k in 1999 and $350k in 2000, whereas the expenditure will be delayed to years 2008 and 2009 for the other two alternatives.

8.5.3.2 Operation Cost. The alternative of retiring the unit basically will not make the total fuel cost of the system be reduced since all the loads in the region have to be supplied from the grid system anyhow after the unit's retirement. For simplicity, the fuel cost has therefore been assumed to be zero for all the three alternatives. Even if the fuel cost were taken into account, it would be the same for the alternatives and make no difference in the comparison.

Maintenance and other operation costs will be zero for the alternative of retiring the unit except for $160k in the first year (1999). It will be $160k/year for the alternative of fixing the unit and $180k/year for the alternative of keeping the unit as is. The latter alternative will require more maintenance activities due to the aging status of the generating unit.

8.5.3.3 Risk Cost. The risk cost can be obtained using the EENS index times a unit interruption cost. In this application, the value of $2k/MWh is used. This is based on the customer damage function of the customer sectors and their compositions in the region. As mentioned earlier, there is no load growth in the region and the risk cost is the same for each year.

The risk cost is $96 \times 2.0 = \$192k$/year for the alternative of retiring the unit, $30 \times 2.0 = \$60k$/year for the alternative of fixing the unit, and $65 \times 2.0 = \$130k$/year for the alternative of keeping the unit as is.

The itemized costs for the three alternatives are summarized in Tables 8.9–8.11. The present values (in 1999 dollars) of the costs are calculated at the discount rate

Table 8.9 Costs for the alternative of retiring the unit (k$)

Year	Capital	Decommission	Fuel	O&M	Risk	Total
1999	0	100	0	160	192	
2000	0	350	0	0	192	
2001	0	0	0	0	192	
2002	0	0	0	0	192	
2003	0	0	0	0	192	
2004	0	0	0	0	192	
2005	0	0	0	0	192	
2006	0	0	0	0	192	
2007	0	0	0	0	192	
2008	0	0	0	0	192	
2009	0	0	0	0	192	
PV (1999)	0	424.1	0	160	1480.3	2064.4

Table 8.10   Costs for the alternative of fixing the unit (k$)

Year	Capital	Decommission	Fuel	O&M	Risk	Total
1999	300	0	0	160	60	
2000	900	0	0	160	60	
2001	0	0	0	160	60	
2002	0	0	0	160	60	
2003	0	0	0	160	60	
2004	0	0	0	160	60	
2005	0	0	0	160	60	
2006	0	0	0	160	60	
2007	0	0	0	160	60	
2008	0	100	0	160	60	
2009	0	350	0	160	60	
PV (1999)	1133.3	212.1	0	1233.6	462.6	3041.6

of 0.08 using Equation (6.4) and are also shown in the tables. It can be seen that the alternative of retiring the unit requires the lowest total cost. The investment and operation/maintenance costs of upgrading the unit capacity to 50 MW make this alternative not cost-efficient, although it can lower the risk cost. The option of "doing nothing" also requires a higher total cost than retiring the unit.

8.5.4 Summary

Based on the results of the overall economic analysis, the option of retiring the local generating unit is competitive since it requires the lowest total cost. According to the risk indices, this alternative also would not lead to significant deterioration in the system reliability compared to the option of keeping its status as is.

Table 8.11   Costs for the alternative of keeping the unit as is (k$)

Year	Capital	Decommission	Fuel	O&M	Risk	Total
1999	0	0	0	180	130	
2000	0	0	0	180	130	
2001	0	0	0	180	130	
2002	0	0	0	180	130	
2003	0	0	0	180	130	
2004	0	0	0	180	130	
2005	0	0	0	180	130	
2006	0	0	0	180	130	
2007	0	0	0	180	130	
2008	0	100	0	180	130	
2009	0	350	0	180	130	
PV (1999)	0	212.1	0	1387.8	1002.3	2602.2

The decision making associated with the retirement of aged equipment has been a challenge for utilities. The risk evaluation and the economic analysis, including the risk cost, provide an effective means to accomplish this.

8.6 CONCLUSIONS

This chapter discusses the application of risk evaluation to generation source planning. It is suggested to perform the risk assessment of generation systems by considering the transmission network constraint instead of using the simple generation–demand model. Introduction of the transmission network into the modeling not only brings the evaluation closer to reality but also makes it possible to investigate the impact of generation source location. The procedure includes the stochastic simulation of the generation cost and risk cost in a composite generation and transmission system and the overall economic analysis, including the risk cost. Conceptually, addition of a generating unit will improve system reliability and thus reduce the system risk cost. The degree of the improvement totally depends on the location and size of added generating units. It is possible that a particular location may only create a very marginal impact on system risk.

There are two assessment methods used in the economic analysis. The first one is to compare the total costs of alternatives, including the investment, operation, and risk costs. This method applies to the case in which a utility is the unique owner of all generation and transmission facilities. The second one is the maximum benefit approach, which is applied in planning the access of cogenerators or independent power producers to a utility system. In the latter case, the investment and operation costs of cogenerating units are not the responsibility of the utility, whereas the access of cogenerators to the system brings both benefits and costs to the utility.

The presented method can be utilized in different generation planning issues, including the comparison among different generation enhancement schemes, identification of the best location and size of cogenerators, and decision making about the retirement of aged generation equipment. It is important to recognize that, as in any other planning process, the risk evaluation and the economic analysis in generation planning are only necessary portions of the whole issue, although they may play a decisive role in the process.

CHAPTER 9

SELECTION OF SUBSTATION CONFIGURATIONS

9.1 INTRODUCTION

Substations and switching stations are critical segments in power systems. They are the points of energy transfers between generation and transmission or transmission and distribution networks. The fundamental principles in selection of substation configurations are reliability, economy, and flexibility in design and operation. Traditionally, the risk evaluation of a substation configuration is conducted separately; the techniques used for such an evaluation have been discussed in Section 5.4. This chapter describes a method to perform the risk assessment for a combinative system consisting of transmission network and substation configurations. The application of the presented method includes the following two cases:

1. The purpose of risk evaluation is to assess the reliability of a substation configuration. The transmission network around the substation is incorporated in the assessment but transmission lines may or may not be assumed to be 100% reliable. In this case, the impact of transmission network constraints on the substation layout, if any, is included in the result.

2. The purpose of risk evaluation is to determine an arrangement of the transmission lines that are directly connected to substation equipment. The transmission lines and substation configurations, which are often called a subtransmission system, are assessed together as a combinative system. In this case, the impact of substation configurations on the transmission arrangement is captured.

Risk Assessment of Power Systems. By Wenyuan Li
ISBN 0-471-63168-X © 2005 the Institute of Electrical and Electronics Engineers, Inc.

Conceptually, the method presented can be used to conduct the risk assessment for a large-scale transmission system with multiple substation configurations incorporated. However, the emphasis in this chapter is placed on the selection of substation configurations or subtransmission arrangements. The load curtailment model for the combinative system of transmission network and substation configurations is illustrated in Section 9.2. The risk evaluation approach and the economic analysis concept are summarized in Section 9.3. Two examples are given in Sections 9.4 and 9.5, respectively. The first one is an application of the approach to the selection of a substation configuration. The second one provides a demonstration of selecting a subtransmission structure with two substations.

9.2   LOAD CURTAILMENT MODEL

The key to the risk evaluation is to assess the load curtailments at load points for each failure state in a combinative system of transmission network and substation configurations. The load curtailment models described in Section 5.5.5 apply to a transmission network but not to a substation configuration. This is because there is no impedance for breaker branches in the configuration and, thus, the breaker branches cannot be incorporated into the Jacobian matrix of power flow equations.

Figure 9.1 shows the skeleton of a combinative system in which one substation in the transmission system has a complete layout diagram (the lower portion of the figure), whereas the others are simplified using single buses (the upper portion of the figure). Depending on the purpose of a study, one or several substations can be represented using a detailed configuration. Generally, if the task is to evaluate the risk to a transmission network by considering the impact of substation configurations, main substations should be modeled in detail. If the task is to evaluate the risk to a substation configuration under the constraint of a transmission network, only the substation configuration under consideration requires a complete representation.

In order to incorporate substation configurations into the model, all buses are classified into four categories:

1. The first one includes the simplified single buses in the transmission network portion, each of which represents a substation.
2. The second one includes the buses to which both transmission lines and substation equipment (breakers or transformers) are connected, such as buses 1, 2, and 3 in the figure.
3. The third one includes the buses inside the substation to which only substation branches (breakers and/or transformers) are connected, such as buses 4, 5, and 6.
4. The fourth one includes the buses to which the substation equipment is connected at one side and a load at the other side, such as buses 7, 8, and 9.

With the bus classification and the expression of power flows on the substation branches, the load curtailment model for transmission network given in Section

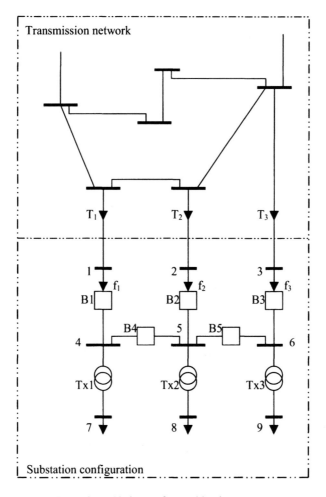

Figure 9.1   Skeleton of a combinative system.

5.5.5 can be extended to include substation configurations. The model is formulated as follows:

$$\min \sum_{i \in ND} W_i C_i \tag{9.1}$$

subject to

$$T_n = \sum_{i \in NB1} A_{ni}(PG_i - PD_i + C_i) - \sum_{k \in NB2} A_{nk}\left(\sum_{i=1}^{N_k} f_i\right) \quad (n = \in LB) \tag{9.2}$$

$$\sum_{i=1}^{N_k} f_i = \sum_{i=1}^{M_k} T_i \quad (k = \in NB2) \tag{9.3}$$

$$\sum_{i=1}^{N_k} f_i = 0 \qquad (k \in NB3) \tag{9.4}$$

$$\sum_{i=1}^{N_k} f_i = C_k - PD_k \qquad (k \in NB4) \tag{9.5}$$

$$\sum_{i \in NG} PG_i + \sum_{i \in ND} C_i = \sum_{i \in ND} PD_i \tag{9.6}$$

$$PG_i^{\min} \le PG_i \le PG_i^{\max} \qquad (i \in NG) \tag{9.7}$$

$$0 \le C_i \le PD_i \qquad (i \in ND) \tag{9.8}$$

$$|T_n| \le T_n^{\max} \qquad (n \in LB) \tag{9.9}$$

$$|f_n| \le f_n^{\max} \qquad (n \in LS) \tag{9.10}$$

where, T_n is the real power flow on branch n in the transmission network and is positive when it flows into a bus; PG_i, PD_i, and C_i are the generation injection, real power load, and load curtailment at bus i, respectively; A_{ni} is the element of the relationship matrix A between the branch real power flows and bus power injections in the transmission network; f_i is the real power flow on branch i in the substation configuration and is positive when it flows out of a bus; $PG_i^{\min}$, $PG_i^{\max}$, $T_n^{\max}$, and $f_n^{\max}$ are the limits of PG_i, T_n, and f_n, respectively; N_k is the number of substation branches that are connected to bus k; M_k is the number of transmission branches that are connected to bus k; NG, ND, and LB are, respectively, the sets of the generation buses, load buses, and branches in the transmission network; LS is the set of branches in the substation configuration; $NB1$, $NB2$, $NB3$, and $NB4$ are, respectively, the sets of the buses in the first, second, third, and fourth category defined earlier; W_i is the weighting factor reflecting the importance of bus loads. The objective of the model is to minimize the total load curtailment while satisfying the power balance, linearized power flow relationship, Kirchhoff's first law at each bus in the substation configuration, and all the limits.

The generation variables are included in the model for generality. These variables may represent real generators if the whole system is modeled or equivalent generation injections at some buses if a reduced transmission system is considered. In the latter case, the equivalent generation variables should be selected at main substations so that a sufficiently large upper limit can be designated.

If a branch in the transmission network fails, a zero branch limit is applied and a sufficiently large branch reactance value is used in calculating the relationship matrix A for the system failure state. If a branch in the substation configuration fails, only a zero branch limit is applied.

9.3 RISK EVALUATION APPROACH

9.3.1 Component Failure Models

The components (lines, cables, or transformers) in the transmission network portion are modeled using two-state (up and down) random variables. Generally, the failures of generators or equivalent generators are not considered since the focus is placed on the risk evaluation of transmission network and substation configurations.

Modeling the failures of components in the substation portion requires more sophisticated representations. Both the active and passive failure modes should be considered. An open-circuit failure of substation components can be modeled in the same way as the transmission components. A bus failure in the substation can be represented using the group outage model described in Section 2.3.2. For example, the failure of bus 4 in Figure 9.1 causes the simultaneous outage of the bus itself together with the two breakers that are directly connected to it, and these components cannot be recovered to the normal state until bus 4 is repaired. A short-circuit failure on breakers or transformers is represented using the switching model given in Section 2.3.3. For example, if the breaker B4 in Figure 9.1 has a short-circuit fault, B1, B2, and B5 are opened, resulting in the switching state in which B1, B2, B4, and B5 plus the transformers Tx1 and Tx2 are out of service. After B4 is isolated by the switches at both its sides, B1, B2, and B5 are reclosed to restore the supply to Tx1 and Tx2, leading to the repairing state with only B4 out of service. B4 is reclosed to get back to the normal state once it is repaired.

9.3.2 Procedure of Risk Evaluation

The procedure of risk evaluation for the combinative system of transmission network and substation configurations includes the following steps:

1. Select a load level in a multiple-level model.
2. Select a system state using the state enumeration or Monte Carlo simulation method. In the selection, the component failure models discussed in Section 9.3.1 are used.
3. Check to see if there is any load bus in the substation configurations that is disconnected from the system, using the connectivity identification method described in Section 5.4.2.
4. Conduct the contingency analysis for the transmission network portion using the method given in Section 5.5.4 to examine whether or not there is any system problem.
5. Solve the load curtailment model in Section 9.2 for the selected system state to minimize the total load curtailment while satisfying all the constraints.
6. Update the risk indices.
7. Repeat Steps 2 to 6 until the convergence rule is met.
8. Go to Step 1 until all the load levels are considered.

It is possible to perform simplifications in some cases. For example, if the task is the risk evaluation of a substation configuration under the constraint of a transmission network, it is generally unnecessary to include the whole transmission network. Incorporating only the transmission lines around the substation into the model is often good enough. If the task is to conduct the comparison between different substation layouts and the failures of some components in the transmission network are estimated to have the same impact, it is normally acceptable to assume that these components are 100% reliable. In this case, the effect of the transmission network in the evaluation is focused on the constraint due to the power-flow equations but not the transmission component failures. If a subtransmission system has such a simple structure that the load curtailments at load points in all failure states can be determined by checking the connectivity, there is no need to solve the minimization model of load curtailment. In this case, the average load model at each individual bus over a time period is generally as accurate as a load curve model.

9.3.3 Economic Analysis Method

Similar to the applications in transmission or generation source planning, the risk evaluation is only one portion of the whole process of selecting substation configurations. A comprehensive economic analysis is always crucial in decision making. The concepts and methods associated with the minimum total cost and present value that have been discussed in Chapters 6 and 8 are still applicable for substation planning. The risk cost due to the unreliability in a substation configuration is one component of the total cost.

9.4 EXAMPLE 1: SELECTING SUBSTATION CONFIGURATION

This example demonstrates an application of the presented approach in selection of substation configurations. The Monte Carlo simulation method is utilized to determine system states.

9.4.1 Two Substation Configurations

The purpose of the study is to compare two substation configurations. The single-line diagrams of the two substation arrangements under the same transmission network environment are shown in Figures 9.2 and 9.3. The 138 kV transmission network has been reduced in such a way that only the transmission lines around the substation are included and the two generators in the figures are the equivalent power injections. Note that the breakers on the transmission lines and the switches on both sides of each substation breaker or transformer are not shown in the figures. The first substation configuration is a traditional section-bus design, whereas the second one is a ring-bus arrangement. Intuitively, the ring-bus structure is more re-

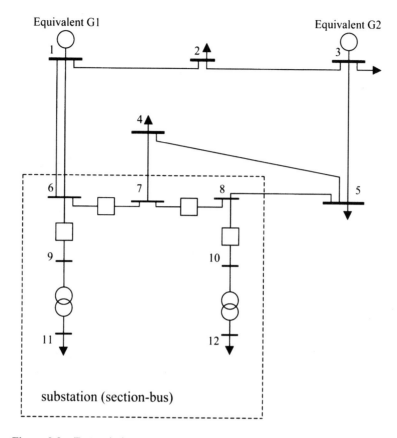

Figure 9.2   Transmission network and section-bus configuration at a substation.

liable but needs one more breaker compared to the section-bus design. The objective of the study is to quantify the difference in risk between the two layouts and examine whether or not the additional breaker in the ring-bus structure can be justified.

9.4.2 Risk Evaluation

9.4.2.1 *Study Conditions and Data.* Both the transmission network and substation configurations are included in the evaluation. The failures of the components are modeled as follows:

- The failures of the transmission lines are represented using the two-state model (up and down states).
- The failures of the breakers and transformers are represented using the three-state model (up, switching, and repair states). In other words, the protection schemes are considered.

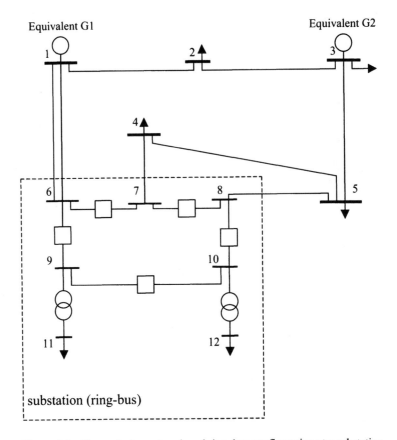

Figure 9.3 Transmission network and ring–bus configuration at a substation.

- The failures of all the busbars in the substation are represented using the group outage model.
- The two equivalent generators are considered to be 100% reliable.

The data of the transmission lines and substation components are shown in Tables 9.1 and 9.2. The bus peak loads in 2002 are given in Table 9.3. The failure data are based on historical statistics. The 12-level load curve model in Table 9.4 is obtained from the average chronological hourly load records in the local area.

9.4.2.2 Results. The following three annual risk indices were evaluated:

1. PLC   Probability of load curtailment
2. EFLC   Expected frequency of load curtailment (failures/yr)
3. EENS   Expected energy not supplied (MWh/yr)

Table 9.1 Data of transmission lines

Line		Capacity	Impedance*		Failure data	
					Frequency	Repair time
From bus	To bus	(MVA)	R (p.u.)	X (p.u.)	(failures/yr)	(h/failure)
1	2	200	0.0564	0.2014	0.454	20.45
1	6	200	0.0493	0.1762	0.301	24.61
1	6	200	0.0493	0.1762	0.301	24.61
2	3	200	0.0502	0.1791	0.272	21.40
3	5	200	0.0429	0.1480	0.541	40.80
4	5	150	0.0170	0.0601	0.267	25.02
4	7	150	0.0202	0.0819	0.311	35.01
5	8	150	0.0301	0.1102	0.281	30.08

*p.u. is based on the 100 MVA and 138 kV base.

Table 9.2 Data of substation components

Component	Capacity (MVA)	Impedance*		Failure frequency		Repair time (h/failure)
		R (p.u.)	X (p.u.)	Active (failures/yr)	Passive (failures/yr)	
Transformer	75	0.0096	0.1914	0.0352	0.0033	500.0
Breaker[†]		0.0000	0.0000	0.0163	0.0012	100.0
Busbar				0.1752		40.0

Switching time = 1 hour

*p.u. is based on the 100 MVA and 138 kV base.
[†]Breakers are assumed to have a sufficiently large capacity.

Table 9.3 Bus peak loads

Bus	Peak load (MW)
2	21.43
3	33.60
4	25.15
5	41.07
11	47.60
12	42.16

Table 9.4   The twelve-level load curve model

Load level (p.u)	Probability
1.00	0.00649
0.95	0.02664
0.90	0.06193
0.85	0.08288
0.80	0.10439
0.75	0.14572
0.70	0.18727
0.65	0.15653
0.60	0.09233
0.55	0.07218
0.50	0.06364
0.45	0.00000

The risk indices of buses 11 and 12 at the low-voltage side of the substation can be used for the comparison between the two substation configurations and are summarized in Table 9.5. The indices of the other buses in the transmission network portion are given in Table 9.6. It should be appreciated that the PLC and EFLC indices of the buses cannot be directly summed up since load curtailment events at different buses are not mutually exclusive. The "total" of the PLC and EFLC indices in Table 9.5 is the sum of the indices at buses 11 and 12 minus the portion contributed by the simultaneous load curtailment events at the two buses. In this example, the common portion is very small indicating that the two buses are dominated by the separate load curtailment events. Unlike the PLC and EFLC indices, the EENS indices at the buses can be summed up to obtain the total EENS index.

The following observations can be made from the results:

- Compared to the section-bus configuration, the ring-bus layout reduces the risk by 24% in the PLC and EENS indices and by 41% in the EFLC index.

Table 9.5   Risk indices at buses 11 and 12

Bus	PLC	EFLC (failures/yr)	EENS (MWh/yr)
Section-bus configuration			
11	0.004102	0.42563	1215.3
12	0.004172	0.43011	1094.8
Total	0.008223	0.85292	2310.1
Ring-bus configuration			
11	0.003132	0.25121	924.7
12	0.003178	0.25332	834.0
Total	0.006250	0.50255	1758.7

Table 9.6 Risk indices at other buses

Bus	PLC	EFLC (failures/yr)	EENS (MWh/yr)
Section-bus configuration			
2	0.0000007	0.00064	0.099
3	0.0	0.0	0.0
4	0.0000009	0.00059	0.148
5	0.0	0.0	0.0
Ring-bus configuration			
2	0.0000007	0.00061	0.094
3	0.0	0.0	0.0
4	0.0000009	0.00058	0.144
5	0.0	0.0	0.0

- In the example, the transmission network does not have a significant impact on the substation configuration. The substation configuration is the symmetric structure connected to buses 11 and 12, although the transmission network provides an asymmetric outside environment. The PLC and EFLC indices at the two buses are basically the same. Note that the difference in the EENS index is due to the different megawatt loads at the two buses.

- The load curtailments at the other buses that are only affected by the transmission network are either extremely small or effectively zero. The load curtailments at buses 2 and 4 are basically caused by the double-line contingency events that have a very low probability of occurrence. Both buses 3 and 5 have no load curtailment, for different reasons. Bus 3 is directly connected to the 100% reliable generation source, whereas the triple-line outage that can lead to the load curtailment at bus 5 was not sampled in the Monte Carlo simulation because of its extremely small probability.

9.4.3 Economic Analysis

There are two cost components: investment and risk costs. The economic analysis can be performed using a benefit/cost assessment. The section-bus configuration is thought of as the base structure. In contrast, the ring-bus configuration needs one more breaker but leads to reduction in the EENS index, which can be converted to the risk cost reduction by multiplying by a unit interruption cost. The planning time span over the 10 years from 2002 to 2011 was considered. The economic data is given in Table 9.7. According to the load forecast, the loads at buses 11 and 12 would be unchanged until 2005, increased by 2% in 2006, and then kept steady until 2011. The unit interruption cost was estimated to be $1.2/kWh. The equivalent annual investment of the additional breaker needed for the ring-bus option, the EENS reduction due to the use of the ring-bus configuration, and the benefit (reduction in the risk cost) from 2002 to 2011 are given in Table 9.8. The equivalent annu-

Table 9.7   Economic data

Investment for a breaker	$770,000
Economic life	45 years
Discount rate	8%

al investment is obtained using Equation (6.2). The present values (PV, in 2002 dollars) of the equivalent annual investment costs and benefits from 2002 to 2011 are calculated using Equation (6.4) and are also shown in the table.

The benefit/cost ratio of using the ring-bus layout versus the section-bus configuration is:

$$4843.56/460.85 = 10.5$$

Obviously, the ring-bus configuration can be financially justified.

9.4.4 Summary

A ring-bus layout is generally more reliable than the traditional section-bus structure but may require more breakers. The financial justification on additional breakers is always a challenging issue in the substation design. The presented risk evaluation method provides a vehicle to perform the probabilistic economic analysis. In the given example, the additional breaker is financially justifiable. This does not mean that a ring-bus arrangement can be justified in every situation. Each case has to be evaluated individually. The method can be also applied to the comparisons among any other substation configurations.

Table 9.8   The equivalent annual investment, the reduction in the EENS, and the benefit due to use of ring-bus configuration

Year	Annual investment (k$/yr)	Reduction in EENS (MWh/yr)	Reduction in risk cost (k$/yr)
2002	63.59	551.4	661.68
2003	63.59	551.4	661.68
2004	63.59	551.4	661.68
2005	63.59	551.4	661.68
2006	63.59	562.4	674.88
2007	63.59	562.4	674.88
2008	63.59	562.4	674.88
2009	63.59	562.4	674.88
2010	63.59	562.4	674.88
2011	63.59	562.4	674.88
PV	460.85		4843.56

The impact of transmission network environments on the risk of substation layouts does not manifest in this particular example. However, it should be kept in mind that this sort of impact often exists. If a determinate judgment cannot be made, it is preferable to conduct the risk assessment on the combinative system, including the transmission network and substation configurations.

9.5 EXAMPLE 2: SELECTING TRANSMISSION LINE ARRANGEMENT ASSOCIATED WITH SUBSTATIONS [50]

This section provides an application of the presented method to determination of a transmission line arrangement supplying two substations.

9.5.1 Description of Two Options

In transmission systems below 230 kV, the tapping connection from a transmission line is a commonly used arrangement for supplying substations along a right of way. The objective of the application is to conduct the comparison between two subtransmission system structures. The two transmission line arrangements with the substation configurations are shown in Figures 9.4 and 9.5. The switches at both sides of the breakers or transformers are omitted in the figures. The transmission line is at the 60 kV level and the low-voltage side is at 12 kV. The two substations, which are located one-third from each end of the transmission line, have the identical layout and each of the four customers has the same amount of load demand. In other words, the two subtransmission systems are identical except that the one is a

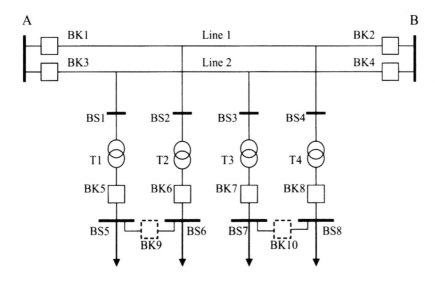

Figure 9.4 Double-line arrangement.

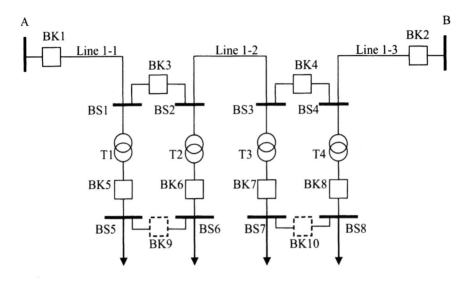

Figure 9.5 Single-line arrangement.

double-line structure and the other is a single-line structure. In the single-line arrangement, the second transmission line is removed and the breakers BK3 and BK4 on the two ends of the second line are moved to the two substations. The two arrangements require the same number of breakers and transformers. At first glance, the double-line option looks more reliable. With a qualitative analysis of switching logic, however, it has been found difficult to make such a judgement.

9.5.2 Risk Evaluation and Economic Analysis

In this example, there is no need to perform any calculations of power flow and load curtailment minimization since the load curtailment in any system state with component outages can be identified directly by means of connectivity checking. An enumeration technique was used to conduct the risk evaluation.

9.5.2.1 *Study Condition and Data.* The study conditions were as follows:

- The up-and-down, two-state model is used for the transmission lines, the three-state model with a switching state is used for the breakers and transformers, and the group outage model is used for the buses.
- The power source points A and B at the two ends of the transmission line arrangements are assumed to be 100% reliable.
- The failures of the buses (BS5, BS6, BS7, and BS8) to which the loads are directly connected are not considered since their impacts on the risk level of the two transmission line arrangements are identical.

- The breakers BK9 and BK10 are normally opened and each one can be automatically closed when either of the breakers at its two sides is opened due to the protection scheme.

The data of the transmission lines, substation components, and loads are shown in Tables 9.9 to 9.11. The length of lines 1 and 2 in the double-line arrangement is 30 km, whereas in the single-line arrangement line 1 is split into the three sections of lines 1-1, 1-2, and 1-3, with equal length of 10 km, and line 2 is removed. The failure data are based on the average historical statistics of the 60 kV facilities. The four customers have equal loads with the same loads factor.

9.5.2.2 Results. The three risk indices of PLC (probability of load curtailment), EFLC (expected frequency of load curtailment), and EENS (expected energy not

Table 9.9 Data of transmission lines

Line	Capacity (MVA)	Length (km)	Failure data	
			Frequency (failures/yr)	Repair time (h/failure)
Line 1	100	30	0.3612	62.5
Line 2	100	30	0.3612	62.5
Line 1-1	100	10	0.1204	62.5
Line 1-2	100	10	0.1204	62.5
Line 1-3	100	10	0.1204	62.5

Table 9.10 Data of substation components

Component	Capacity (MVA)	Failure frequency		Repair time (h/failure)
		Active (failures/yr)	Passive (failures/yr)	
Transformer	40	0.0761	0.0026	490.0
Breaker		0.0124	0.0012	80.4
Busbar		0.0912		25.2

Switching time = 1 hr

Table 9.11 Data of bus loads

Bus	Peak load (MW)	Load factor
BS5	14	0.7143
BS6	14	0.7143
BS7	14	0.7143
BS8	14	0.7143

supplied) were evaluated. The results for the two transmission line arrangements are shown in Table 9.12. These are the total indices for the four load points. Again, note that the total PLC or EFLC index is not the direct sum of the indices at the load points. The total PLC or EFLC is the probability or frequency of load curtailments happening at least at one load point. Each index is divided into the two portions. The first one is associated with the load curtailments in the switching states that exist only in a short period. The second one is associated with the load curtailments in the repairing states that would last until the end of the repair process, a relatively long period.

The following observations can be made:

- The single-line arrangement has a lower probability of load curtailments and a smaller expected energy not supplied, but a higher expected frequency of load curtailments compared to the double-line arrangement. By looking at the composition of the indices, all the PLC, EFLC, and EENS indices due to the repairing states for the single-line arrangement are smaller than those for the double-line arrangement. Although the indices due to the switching states for the single-line arrangement are larger than those for the double-line arrangement, this portion of the indices is only associated with one hour of switching time.

- The relatively high frequency of load curtailments in the single-line arrangement is due to failure events that can be cleared by switching actions.

- The EFLC indices are dominated by the portion of the switching states that is of short duration, whereas the PLC and EENS indices are dominated by the portion of the repairing states that is of long duration.

- Generally, the EENS index carries more information than the PLC and EFLC indices since it is a combination of the duration, frequency, and consequence (amount of load curtailments) of failure events. The total EENS for the single-line arrangement is about 60% of that for the double-line arrangement.

Table 9.12   Risk indices

Arrangement	PLC	EFLC (failures/yr)	EENS (MWh/yr)
Double-line arrangement			
Switching states	0.000005748	0.050896	1.0540
Repairing states	0.000107392	0.016442	20.7702
Total	0.000113140	0.067338	21.8242
Single-line arrangement			
Switching states	0.000009164	0.081147	1.1851
Repairing states	0.000066006	0.008531	11.8343
Total	0.000075170	0.089678	13.0194

It can be concluded from the quantified risk evaluation that the single-line arrangement is more reliable than the double-line arrangement in this example.

9.5.2.3 *Economic Analysis.* In this particular example, the economic analysis is obvious and straightforward. The single-line arrangement not only saves one transmission line of 30 km but also provides a lower power supply risk in comparison to the double-line arrangement.

9.5.3 Summary

In the example, the single-line arrangement is superior to the double-line arrangement not only in an economic sense but also in accordance with the risk level. However, it is important to appreciate that this conclusion cannot be generalized. The application just provides an example indicating that a double-line arrangement is not necessarily more reliable than a single-line arrangement, particularly when a protection scheme is adopted.

The lower PLC and EENS indices for the single-line arrangement are due to more switching possibilities. It should be recognized that switching can reduce the total probability of load curtailments and the expected energy not supplied by shortening the duration of load curtailment states, but cannot reduce the frequency of failure events.

9.6 CONCLUSIONS

This chapter discusses the application of risk evaluation in determining subtransmission system structures, which includes selection of substation configurations or transmission line arrangements associated with substation layouts. The approach to performing the risk assessment for the combinative system of transmission network and substation configurations has been presented. Generally, evaluating substation configurations under the constraint of a transmission network provides more accurate results than evaluating only substation configurations since the transmission network and failures of its components may have impacts on the reliability of substation configurations. On the other hand, substation layouts may also affect the reliability of a transmission network, particularly the transmission line arrangements that are directly connected to the substations through tapping connections.

In implementation of the presented approach, it is possible to consider simplifications depending on the case. For instance, instead of the whole transmission network, only the transmission lines around the substation configurations under study may be included in the evaluation. The component failures that have the same impact on the arrangements in comparison can be excluded from the assessment. In some cases, if load curtailments at load points can be identified through checking connectivity, the process of solving the minimization model of load curtailments, which is always associated with a calculation burden, can be avoided.

In the risk evaluation of substation configurations, the switching actions due to protection schemes can generally provide improvements in the probability and energy-related risk indices. However, they cannot reduce or even may increase the frequency of failure events. This may lead to the case in which one alternative has the better probability or energy-related index and another one has the better frequency index. In such a situation, an engineering judgment is necessary. In general, the EENS index should outweigh the other indices since it is a combination of frequency, duration, and severity of consequences.

As in the application to transmission or generation source planning, the economic analysis, including the risk cost, is imperative in substation planning. The minimum total cost method or the benefit/cost analysis technique is the same as that used in Chapters 6 and 8.

CHAPTER 10

RELIABILITY-CENTERED MAINTENANCE

10.1 INTRODUCTION

Maintenance is one of the major activities in electric utilities. It includes the regular field assessment, overhaul, refurbishment, and replacement of equipment. There are two types of maintenance: preventive and corrective. Traditional maintenance focuses on equipment itself, including investigations into its physical condition, operation performance, and field environment. An important fact, which has been more or less ignored in traditional maintenance, is that taking a component out of service for maintenance is always associated with a potential increase in the operation risk of the whole system during the maintenance period. To include this factor in the decision-making process, quantitative risk evaluation is required to identify the impact of a maintenance outage on the reliability of the whole system. One fundamental principle in reliability-centered maintenance (RCM) is that the importance of a component and its maintenance strategy do not mainly depend on the component itself but on its effects on system reliability. At the same time, it should be appreciated that it is basically impossible to incorporate all aspects of a maintenance plan into a single reliability model. The assessments that have been performed in traditional maintenance, such as physical conditions of components, their failure history and aging status, safety concerns in maintenance, workforce limitations and environment impacts, must still be covered in the process. In other words, a reliability-centered maintenance approach is an enhancement to but not a replacement for the traditional maintenance method.

Reliability-centered maintenance has a wide range of implications. The discussion presented here may be quite different from the others that use the same term. Individual cases require different risk evaluation techniques. These techniques have

Risk Assessment of Power Systems. By Wenyuan Li
ISBN 0-471-63168-X © 2005 the Institute of Electrical and Electronics Engineers, Inc.

been discussed in Chapter 5 and are not repeated. The intent of this chapter is to focus on the general concepts and procedures in reliability-centered maintenance.
Section 10.2 describes the four basic tasks that can be performed in RCM. Three
examples are given in Sections 10.3, 10.4, and 10.5 that demonstrate the particulars
in different application cases.

10.2   BASIC TASKS IN RCM

This section presents the general procedure of several basic tasks in reliability-centered maintenance. Each task has a different objective and somewhat different procedure. It should be pointed out that the procedure in each task given below is only
an outline and the details in each step can be worked out with the knowledge gained
from the previous chapters. Some special aspects may need to be further developed
for a particular application. The three examples given later will provide more information.

10.2.1   Comparison between Maintenance Alternatives

The most popular case is to compare different maintenance alternatives. An alternative refers to a maintenance scheme for a single component or a maintenance sequence associated with multiple components. The procedure includes the following
steps:

Step 1: Select the alternatives for comparison based on the assessment of equipment physical conditions and the constraints in implementation of maintenance.

Step 2: Build the risk evaluation model used for the system that the component
to be maintained is a part of. Obviously, the model should be different
for generation, transmission, substation, or distribution systems. The
common point is that any component to be maintained is represented using a determinate planned outage.

Step 3: Evaluate the risks of all the selected alternatives. The risk evaluation
methods have been discussed in Chapter 5. It is worth mentioning that it
is better to use a risk index like the EENS (expected energy not supplied), which is a combination of probability and consequence. If it is
difficult to calculate such an index for a particular case, the "pure" probability index could be used.

Step 4: If necessary, conduct an economic analysis. This requires estimation of
the total cost, including the maintenance cost and the increased system
risk cost due to the maintenance outage.

Step 5: Determine the best maintenance alternative according to a criterion such
as the lowest system risk or the lowest total cost.

It can be seen that the procedure is similar to that used in probabilistic system development planning (see Chapter 6). The difference is that the alternatives in reliability-centered maintenance are associated with the planned outage of one or more system components, not with the addition of new components as in development planning.

10.2.2 Lowest-Risk Maintenance Scheduling

Determination of the timing and duration of a maintenance activity is another task of reliability-centered maintenance. There are many time-dependent factors in power system operation, such as constantly varying loads, generation patterns, reservoir situations, and line ratings in different seasons. This leads to the fact that maintenance outages of the same equipment at different times will have varied impacts on system operation risk. The procedure to determine the lowest-risk maintenance scheduling is as follows:

Step 1: Build the risk evaluation model for the system in which the maintenance activity is carried out. The model must be able to simulate the time-dependent factors.

Step 2: Evaluate the risk to the system with the maintenance outage shifted over all possible time intervals. Generally, the Monte Carlo method is more suitable than the enumeration technique in this case.

Step 3: Determine the lowest-risk maintenance schedule by comparing the results.

Section 10.3 will demonstrate the details using transmission maintenance scheduling as an example.

10.2.3 Predictive Maintenance Versus Corrective Maintenance

In general, the preventive overhaul policy is applied to major equipment in power systems, whereas the corrective repair policy is applied to less important devices. However, random failures of the major equipment cannot be fully avoided through preventive maintenance activities, and corrective repairs are still required. The preventive overhauls may be delayed due to the restriction in the maintenance budget. This results in the increased likelihood of corrective maintenance. The spare policy is a usual strategy for some equipment. In principle, the components in the same type that share one or more spares do not undergo a preventive overhaul before they fail. The preventive maintenance requires a higher annual budget, whereas the corrective maintenance results in a higher failure risk. The comparison analysis between them is an important topic in reliability-centered maintenance. The procedure to conduct the comparison is summarized as follows:

Step 1: Estimate the improvement in the failure parameters that could be contributed by a preventive maintenance activity versus that of a corrective maintenance policy. This includes the possible decrease in the failure frequency and/or repair time, or the improvement of the parameters in the aging-failure model. This step is strongly related to the engineering judgment of experienced engineers.

Step 2: Build the risk evaluation model for the system in which the preventive or corrective maintenance activity takes place.

Step 3: Evaluate the system risks for the base system and the preventive and corrective maintenance cases.

Step 4: Perform the benefit/cost analysis. For the preventive maintenance, the benefit is the reduction in the system risk due to the decreased component failure frequency, repair time, or unavailability, whereas the cost is the budget required. For the corrective maintenance, the benefit is the saving in the maintenance, whereas while the cost is due to the increased system risk and consequences after the failure (longer repair time and/or more severe damage).

Step 5: Compare the results based on the benefit/cost ratios.

10.2.4 Ranking the Importance of Components

A ranking list of component importance is extremely useful information in preparing a strategic maintenance plan. One basic concept of reliability economics is that the value of a component depends on the damage caused by its absence from the system. In other words, a component with a higher capital investment does not have to be more valuable to system reliability than a component with a lower capital investment. It is necessary to rank the importance of components in terms of the impact caused by their absence from the system. The impact is varied in the different system states that are associated with other component forced outages and can be quantified using the risk index, which is a combination of probabilities and consequences of all possible system states. The following is the procedure:

Step 1: Build a system risk assessment model.

Step 2: Evaluate the system risk of the base case in which all components are in service but can randomly fail.

Step 3: Evaluate the system risk of the case in which a component is out of service and all the other components are in service but can randomly fail.

Step 4: Repeat Step 3 for all the components that are considered for ranking.

Step 5: Calculate the differences in the risk indices between the base case and the cases with each component out of service.

Step 6: Create the ranking list of component importance in the light of their impacts on the system risk.

10.3 EXAMPLE 1: TRANSMISSION MAINTENANCE SCHEDULING [23]

Transmission maintenance planning includes two tasks: (1) selecting the most reliable operation mode during the maintenance outage and (2) determining the lowest-risk maintenance schedule. The general procedure of transmission reliability-centered maintenance is discussed first and an actual example is presented that focuses only on the second task—maintenance scheduling—since the first task has been addressed in Chapter 7.

10.3.1 Procedure of Transmission Reliability-Centered Maintenance

Transmission reliability-centered maintenance includes the following four steps.

1. Conduct the traditional maintenance planning to determine a preliminary plan that indicates the component(s) due for a maintenance outage, the available period to do the maintenance (for example, from April to June), the time length needed for the maintenance (for instance, two weeks), and other conditions. In this step, the following factors are considered:
   - Safety code. This can include the conductor-to-ground clearance, working environment evaluation, right-of-way permit, aerial safety patrol, possible limitations due to by-laws, and so on.
   - Investigations into the physical component situation, failure history, age of components, operation history against design conditions, and so on. The investigations assist in establishing a priority list for maintenance and also provide the data for performing the quantified risk evaluations in Steps 3 and 4.
   - Workforce planning
2. Perform the system analysis to select feasible system operation modes during the maintenance outage. This step is generally associated with power-flow calculations, contingency analyses, and voltage and transient stability assessments. The purpose of Step 2 is to determine if the preliminary plan in Step 1 is acceptable or should be modified due to system operational constraints. For instance, if Step 1 suggests that the available period for the maintenance outage is between April and June but there are some security concerns (such as operational limits) or economic considerations (such as a high power export) in April, the period would be adjusted to the one between May and June. If the maintenance outage has to take place as soon as possible because of a safety or environmental condition, additional consequent operation schemes may have to be investigated to overcome the difficulty in the operation or to take a higher but still tolerable system operation risk. Therefore, there must be coordination between the two steps.

In most cases, more than one acceptable system operation mode can be identified for the maintenance outage after the coordination in the above two steps.

3. Select the most reliable operation mode from the several acceptable ones that have been identified in Step 2 by using the risk evaluation method for transmission systems.

4. Determine the lowest-risk maintenance schedule within the period available for the maintenance outage by using the time-shift-based transmission system risk evaluation method.

The reliability-centered maintenance approach presented here is just an extension of the traditional maintenance planing since it includes all the activities required in the latter. The purpose of the last two steps is to obtain the lowest-risk maintenance plan by quantifying the impact of the maintenance outage on the whole system risk within the period available for the maintenance. The basic considerations in the risk evaluation are as follows:

- In the evaluation, the system state with the maintenance outage is treated as the starting point or base case and the combinations of all possible subsequent failures of other system components are simulated. During the maintenance outage, the other system components can randomly fail. Generally, the impacts of forced failures are expected to be much more severe in the system states with the maintenance outage than in the system states without it. The difference is the increased system risk caused by the maintenance. The Monte Carlo sampling technique can be used to simulate the random forced failures of all the other system components. In the case in which there exists the mutual impact between generation patterns and the maintenance outage, the forced failures of both transmission components and generating units have to be simulated. Otherwise, only the failures of transmission components are considered.

- After a forced failure system state is selected in Monte Carlo sampling, it is necessary to evaluate the impacts of failure components on the system risk. This failure state may or may not have a negative effect on system operation, depending on the number of failed components and their locations. Most utility systems are designed to meet the single-contingency criterion, meaning that any one single failure will not create a system problem. However, during the maintenance outage, we already have at least one component out of service and, thus, a forced outage is equivalent to a double outage, two forced outages to a triple outage, and so on.

- The duration of a maintenance outage is generally shorter than the period available to perform the maintenance. For example, the duration may be only two weeks and the maintenance can be performed in the period from April to June. To determine the lowest-risk maintenance schedule, the Monte Carlo simulation of the duration needed by the maintenance outage must be shifted within the period available. The shifted duration may include the time intervals associated with different system operation parameters and patterns. For instance, there are different capacity ratings of transmission lines, cables, and transformers for the summer and winter months, and generation patterns may be changed in different months.

It can be seen that the risk evaluation method described in Section 7.3 meets the above modeling considerations if a time-shifting Monte Carlo simulation process can be appropriately handled. It is not difficult to automate the shifting simulation as long as the Monte Carlo method is designed to have the capacity of dealing with any time length of load curve. The details of risk evaluation are, therefore, not discussed here.

10.3.2 Description of the System and Maintenance Outage

The example is an actual application that was performed in 2001. The system was the 230 kV and 60 kV north metro system of the utility shown in the upper portion of Figure 7.4. There were eight 230 kV underground cables in the north metro system and two of them reached their end-of-life stage. The cable portions of the circuits 2L39 and 2L40 had served for more than 45 years and the field assessment revealed serious physical deterioration. Based on the field assessment result, failure history, and aging condition, as well as safety and environment considerations, it was decided to replace the two cables. Identifying the most reliable operation mode during circuit 2L39 replacement has been discussed in Section 7.4 as the application to the transmission risk evaluation in operation planning. This example focuses on determination of the lowest-risk schedule of circuit 2L40 replacement. Selection of the most reliable operation mode during 2L40 replacement is similar to the case of 2L39 and is not discussed. The study conditions included:

- The new cable of 2L39 was in service and the 2L40 replacement was estimated to take about 6 months to complete.
- The whole year period from November 2001 to October 2002 was available for the implementation of the replacement. The months from October to March were assumed to be in the winter and those from April to June in the summer.
- The different line/transformer ratings and local generation patterns for the winter and summer periods were considered. Particularly, the dynamic ratings of all the cables were applied.
- The outage data of lines, cables, and transformers were based on the failure statistics of the previous 15 years. The aging failures of all the cables were also considered.
- The terminal-related line outages and dependent failures of line sections were modeled.
- The 8760 hourly system load records in 2000/2001 were used to represent the shape of the annual load curve, whereas the peak load corresponded to the load forecast for 2001/2002.

- The shifting Monte Carlo simulation for the lowest-risk schedule of the 2L40 replacement was based on the most reliable operation mode during the 2L40 replacement that had been selected in the first stage of the project.

10.3.3 The Lowest-Risk Schedule of the Cable Replacement

The time shifting Monte Carlo simulation was conducted to identify the lowest-risk schedule for the 2L40 replacement outage. Generally, if there are two or more components planned for outages at the same time, the shifting time interval should be the minimum overlapping duration. In most cases, however, only one major transmission component in a region is scheduled for maintenance at a time, unless there is an unusual circumstance such as a serious safety or environment concern. In this example, the shifting simulation was conducted on a monthly basis. The time length required by the 2L40 replacement was 6 months. Therefore the simulation was performed over the period from November 2001 to April 2002 first, then shifted by one month, (i.e., from December 2001 to May 2002), and then shifted by another month (from January 2002 to June 2002), and so on.

The EENS indices in the seven shifting periods are given in Table 10.1 and shown in Figure 10.1. It can be seen that the lowest system operation risk period for the 2L40 replacement is from April to September 2002. The system risk during this period is about 45% lower compared to the highest-risk period (November 2001 to April 2002). It is interesting to note that this period is delayed by one month compared to the lowest load profile period, which is from March to August. This fact suggests that the lowest load profile period is not necessarily the lowest-risk duration for a planned outage because the system risk is associated with not only the load level but also other system factors such as line ratings, generation patterns, configurations, and component failure probabilities.

10.3.4 Summary

The lowest-risk maintenance schedule can be identified using the time shift risk evaluation over the period available to the maintenance outage. Although the given example is the transmission cable replacement, the procedure should be the same

Table 10.1 EENS indices in the seven shifting periods

Shifting period	EENS (MWh)
November 2001 to April 2002	2312
December 2001 to May 2002	2161
January 2002 to June 2002	1794
February 2002 to July 2002	1550
March 2002 to August 2002	1331
April 2002 to September 2002	1279
May 2002 to October 2002	1502

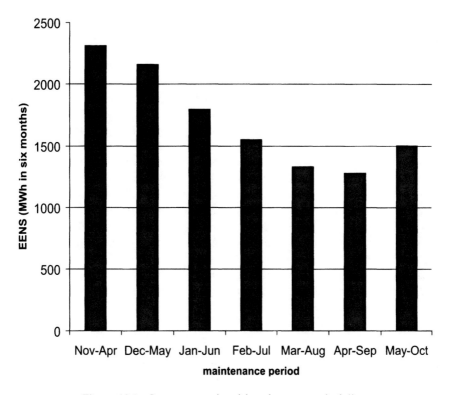

Figure 10.1 Lowest operation risk maintenance scheduling.

for other system components and any maintenance activities. The key is to determine the system model and system conditions so that an appropriate risk assessment technique can be selected.

It should be stressed that the lowest-risk maintenance schedule is only one portion of the whole reliability-centered maintenance scheme. Determining the most reliable operation mode during the maintenance outage is another task. Particularly, the conventional maintenance planning activities, including the physical assessment of the component, safety code, workforce review, are also the imperative part of the proposed RCM process.

10.4 EXAMPLE 2: WORKFORCE PLANNING IN MAINTENANCE [57]

Traditionally, workforce planning in maintenance is only based on the estimation of the amount of work to be done and manpower available. The reliability-centered maintenance approach enables us to investigate this issue from the angle of system risk and benefit/cost analysis. This example illustrates an application of risk evaluation to workforce planning.

10.4.1   Problem Description

The application is associated with the annual preventive overhaul of the HVDC lines in a utility system. An island region is supplied through two 500 kV AC transmission lines, two 138 kV AC lines, two HVDC poles, and local generators, as shown in Figure 10.2. According to the maintenance plan, the two HVDC poles need a preventive annual overhaul that takes the 107 days in total from March 16 to June 30, which is the off-peak load period during one year. It includes 50 days of maintenance on each pole and 7 days on both the poles. In each of the two 50-day periods, the capacity of one pole will be reduced to half while the other pole can keep its full capacity. In the final 7 days, both the poles have to be shut down. There is no need to shed loads in the normal operation state, even if both the HVDC poles are out of service for maintenance. However, the other components, particularly the 500 kV lines, may randomly fail during the HVDC maintenance and the island supply system would be exposed to a higher operation risk in the maintenance period. On the other hand, it is intuitively known that the risk increase is relatively small because of the large capacity redundancy in the supply system during the off-peak period. The question faced by the utility was, would it be worth putting on more manpower to shorten the duration of the overhaul?

10.4.2   Procedure

It can be seen from Figure 10.2 that this is a generation–demand system since all the transmission lines and HVDC poles can be treated as the "generation" sources in the risk evaluation. The procedure provided here can be conceptually applied to generating unit maintenance planning, although the overhaul in this example is associated with the HVDC poles.

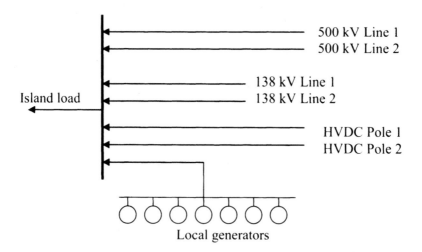

Figure 10.2   Power supply system of an island region.

The procedure includes the following steps:

1. The 107 days from March 16 to June 30 are considered as the time frame in the study. The following maintenance options are prepared for the risk evaluation:

   • Normal maintenance: The overhaul of Pole 1 is conducted for the first 50 days, followed by the overhaul of Pole 2 for the second 50 days, and, finally, the maintenance on both the poles for the last 7 days.

   • Shortened maintenance: One or more modified plans with a shortened duration of maintenance. The risk evaluation for the modified plans must be performed over the same total time length of 107 days so that the results are comparable with those of the normal maintenance. For example, if the duration of the overhaul is shortened to 54 days, the risk evaluation for the other 53 days is still needed, in which all the components (including the HVDC poles) are in service but can randomly fail.

2. The risk evaluation technique described in Section 5.2 is used to assess the risk indices of the island supply system for the maintenance options prepared in Step 1. In the evaluation, the maintenance outages of the HVDC poles are the deterministic events in the given time intervals and all the other components are in service but their random forced failures are simulated.

3. The difference in the EENS (expected energy not supplied) or EDC (expected damage cost) indices between the normal and shortened maintenance options is the benefit due to reducing the duration of the overhaul. The additional cost is the extra manpower required by the shortened maintenance.

4. The benefit/cost analysis is conducted and a decision can be made based on the given criterion.

10.4.3   Case Study and Results

The following three maintenance options were considered:

1. Normal overhaul: the partial Pole 1 outage for the first 50 days (from March 16 to May 4), the partial Pole 2 outage for the next 50 days (from May 5 to June 23), and the outage of both the poles for the last 7 days (from June 24 to June 30)

2. Shortened Overhaul A: the partial Pole 1 outage for the first 25 days (from March 16 to April 9), the partial Pole 2 outage for the next 25 days (from April 10 to May 4), and the outage of both the poles for the last 4 days (from May 5 to May 8)

3. Shortened Overhaul B: the partial Pole 1 outage for the first 25 days (from May 8 to June 1), the partial Pole 2 outage for the next 25 days (from June 2 to June 26), and the outage of both the poles for the last 4 days (from June 27 to June 30)

The assessment was performed for the three consecutive years. The load growth in the island region was estimated to be 1.5% per year. The load-curve shape was based on the hourly load records for the region. The failure data of the AC lines, HVDC poles, and generators were from the historical statistics for the previous 10 years. The composite customer damage function [1] for the island region was used in calculating the EDC index.

The state sampling method described in Section 5.2.2 was used to conduct the risk evaluation for the three annual maintenance options of the HVDC poles. The EENS and EDC indices for the three load level years are shown in Table 10.2. The benefits due to shortening the duration of the HVDC overhaul are the reductions in the EDC index and are listed in Table 10.3. The additional cost required by shortening the overhaul duration is mainly due to the extra manpower and overtime work and was estimated at $80k/year. The average benefit/cost ratios for the two shortened maintenance options of the HVDC poles are calculated as follows:

Shortened Option A: 48.4/80.0 = 0.605
Shortened Option B: 72.2/80.0 = 0.903

The benefit/cost ratios for both the shortened options are smaller than 1.0. The results, therefore, cannot justify their implementation.

10.4.4   Summary

Workforce planning in maintenance is an interesting topic in utilities. The risk evaluation method provides a vehicle to conduct the benefit/cost analysis associated with the workforce. The example is the HVDC line maintenance issue but the risk assessment has been treated as a generation system reliability model because of the problem's nature. The concept can be extended to other similar cases.

Table 10.2   EENS and EDC indices for the three maintenance options in the 107 days

Maintenance option	EENS (MWh)	EDC (k$)
First-year load level		
Normal Option	90.3	172.4
Shortened Option A	66.5	127.0
Shortened Option B	55.2	105.4
Second-year load level		
Normal Option	101.5	193.9
Shortened Option A	76.3	145.7
Shortened Option B	63.7	121.7
Third-year load level		
Normal Option	114.2	218.1
Shortened Option A	87.2	166.6
Shortened Option B	73.6	140.6

Table 10.3   Benefits of the two shortened maintenance options

| | EDC reduction (k$) | | | |
Maintenance option	Year 1	Year 2	Year 3	Average
Shortened Option A	45.4	48.2	51.5	48.4
Shortened Option B	67.0	72.2	77.5	72.2

It should be appreciated that the results are only valid under the study conditions and assumptions. The study must be updated as the time advances. The possible variation in the load growth of the region and in the failure probabilities of the components may lead to a different conclusion. This is a general feature of risk-evaluation-based studies.

10.5   EXAMPLE 3: A SIMPLE CASE PERFORMED BY HAND CALCULATIONS [69]

The purpose of this example is to demonstrate that the reliability-centered maintenance approach can be performed through simple calculations in some cases.

10.5.1   Case Description

The four 500 kV lines—two in parallel in series with the other two in parallel—form a power supply corridor from a generation source center to a load center, as shown in Figure 10.3. The transfer capacity of the lines meets the single-line contingency criterion, that is, any single line outage will not cause any load curtailment.

The maintenance outages of lines 1, 2, and 3 were planned to repair the damages of several towers. 10 days were needed for repairing the towers of line 2 and 5 days for line 1 or line 3. The following two schedules were considered:

1. "Series style": Line 2 is maintained first from September 19 to September 28, followed by line 1 from September 29 to October 3, and then by line 3 from October 4 to October 8.
2. "Parallel style": Line 2 is maintained from September 19 to September 28. The maintenance on line 1 starts on the same date and for the first 5 days

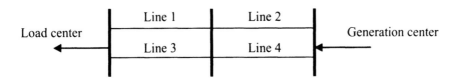

Figure 10.3   A simple power supply system.

(from September 19 to September 23), followed by maintenance on line 3 for the second 5 days (from September 24 to September 28).

The idea is shown in Figure 10.4. The advantage and disadvantage of the two schedules are somewhat complementary and contradictory. Schedule 1 takes twice as long as Schedule 2, meaning that the supply system would be exposed to the risky situation longer. On the other hand, there are fewer forced failure events causing loss of load during the line maintenance outage for Schedule 1 than for Schedule 2. It is difficult to use the traditional criteria to make a judgement about which schedule is better. A simple risk evaluation can answer this question.

10.5.2 Study Conditions and Data

The power source from the generation center is assumed to be 100% reliable and the lump sum load is considered to be at the load center. The study focuses on the impacts of the single line forced-failure events after one or two lines are taken out for maintenance. The double forced contingencies are not considered because the probabilities of their occurrence within 10 or 20 days are extremely low and negligible.

The historical load records of the previous three years indicate that the average hourly load is 1045 MW from September 19 to September 30 and 1242 MW from October 1 to October 8. According to the failure statistics of the previous 10 years, the failure frequency and repair time for each of the four 500 kV lines are obtained and given in Table 10.4. It is interesting to note that the forced outage frequencies for lines 1, 2, and 3 are larger than that for line 4. This fact indicates that the failure frequency is not necessarily proportional to the line length. Generally, if there are sufficient statistics, the failure data of individual lines should be used instead of the average value obtained using the lines at the same kilovolt level.

10.5.3 EENS Evaluation

For each maintenance schedule, there are several single-line forced-failure events causing loss of load during the maintenance outage. For Schedule 1, these are the events:

Event 1a: Line 4 fails while line 2 is out of service for 10 days in September
Event 1b: Line 3 fails while line 1 is out of service for 2 days in September and 3 days in October
Event 1c: Line 1 fails while line 3 is out of service for 5 days in October

For Schedule 2, these are the events:

Event 2a: Line 3 fails while lines 1 and 2 are out of service for 5 days in September
Event 2b: Line 4 fails while lines 1 and 2 are out of service for 5 days in September
Event 2c: Line 1 fails while lines 2 and 3 are out of service for 5 days in September
Event 2d: Line 4 fails while lines 2 and 3 are out of service for 5 days in September

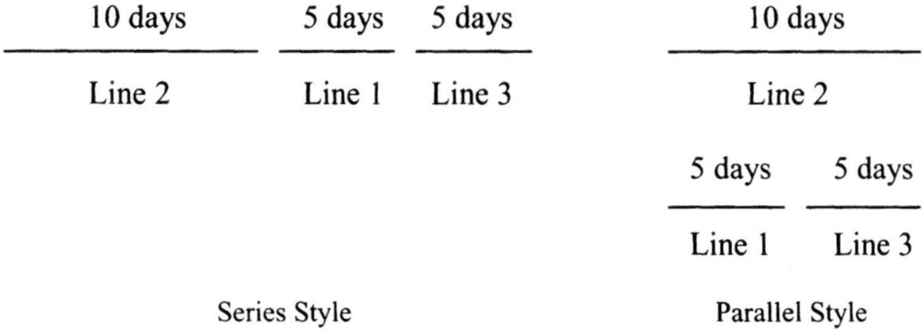

10 days	5 days	5 days	10 days
Line 2	Line 1	Line 3	Line 2

		5 days	5 days
		Line 1	Line 3

Series Style   Parallel Style

Figure 10.4 Two maintenance schedules.

In this application, the comparison between the probabilities of the failure events in the two schedules does not provide the entire picture of the risk caused by the maintenance activities. The risk is associated with not only the probabilities of the failure events but also the variable load profiles during the maintenance. The expected energy not supplied (EENS) is the appropriate index for the risk comparison.

The method of calculating the EENS is explained using event 1a in Schedule 1. The average failure frequency of line 4 during the 10 days of line 2 maintenance is $0.714 \times 10 \times 24/8760$. Once this failure occurs, the average load of 1045 MW (in September) would be curtailed for 0.88 hours. Theoretically, the failure event can take place at any time during the 10 days. If it occurs at some point near the end of the 10-day period, it may last shorter than 0.88 hours due to the recovery of line 2 before the completion of the line 4 repair. However, ignoring this particular case and using the average repair time of 0.88 hours will not create a significant error since the repair time is so much shorter compared to the maintenance duration. The EENS indices for the failure events of the two maintenance schedules are calculated as follows:

Schedule 1:

Event 1a: $(0.88 \times 0.714 \times 10 \times 24/8760) \times 1045 = 17.99$ MWh
Event 1b: $(18.35 \times 1.714 \times 2 \times 24/8760) \times 1045 + (18.35 \times 1.714 \times 3 \times 24/8760) \times 1242 = 501.16$ MWh
Event 1c: $(2.76 \times 1.5 \times 5 \times 24/8760) \times 1242 = 70.44$ MWh
Total: 589.59 MWh

Table 10.4 Failure data of the four lines

Line	Failure Frequency (failures/yr)	Repair time (h/failure)	Length (km)
Line 1	1.50	2.76	31
Line 2	1.857	22.84	80
Line 3	1.714	18.35	38
Line 4	0.714	0.88	80

Schedule 2:

Event 2a: $(18.35 \times 1.714 \times 5 \times 24/8760) \times 1045 = 450.24$ MWh
Event 2b: $(0.88 \times 0.714 \times 5 \times 24/8760) \times 1045 = 8.99$ MWh
Event 2c: $(2.76 \times 1.5 \times 5 \times 24/8760) \times 1045 = 59.26$ MWh
Event 2d: $(0.88 \times 0.714 \times 5 \times 24/8760) \times 1045 = 8.99$ MWh
Total: 527.48 MWh

It can be seen that Schedule 2 is better than Schedule 1 since its total EENS index is about 10.5% lower. It is important to appreciate that it should not be concluded that the parallel maintenance style is always better than the series maintenance style. The conclusion is valid only for the given data. If the load levels and/or failure data are changed, the results may be different.

10.5.4 Summary

The example indicates that even for a very simple case, it may still not be straight-forward to select a maintenance scheme from the qualitative analysis. However, quantified risk evaluation can provide a clear comparison between the mainte-nance alternatives. Reliability-centered maintenance does not have to be always sophisticated and can be performed using the simple risk assessment method in some cases.

10.6 CONCLUSIONS

This chapter discusses the concepts of reliability-centered maintenance in power systems. The focus has been placed on the procedures. Four basic tasks are ad-dressed and three examples are given to demonstrate the details of the different ap-plications. It should be recognized that reliability-centered maintenance has a wide range of implications and is not limited to the concepts and four tasks provided in this chapter.

There are two essential features in the reliability-centered maintenance approach presented in the chapter. First, the key to the approach is to quantitatively assess the impact of a maintenance activity on the whole system risk rather than to focus on the situation of the individual components to be maintained. This is based on the principle in reliability economics that the value of a component depends on the in-creased risk that can be caused by absence of the component from the system. Sec-ond, the reliability-centered maintenance process includes all the traditional mainte-nance planning activities so that it can be easily accepted and implemented in the existing maintenance practice of utilities.

The risk evaluation methods used in the reliability-centered maintenance process can be chosen from those that have been described in Chapter 5. In selecting the method, it is important to determine an appropriate system model. As seen in Ex-ample 2, the generation–demand risk evaluation method can be used for the maintenance of transmission components as long as the issue can be addressed using the generation system model. In a simple case like Example 3, even the manual calculation procedure can be applied. Any maintenance scheduling is always associated with a time period. The time-dependent factors, such as load curves, time-variant line ratings and generation patterns, should be considered in the risk assessment for maintenance outages.

CHAPTER 11

PROBABILISTIC SPARE-EQUIPMENT ANALYSIS

11.1 INTRODUCTION

The sophisticated analysis of spare equipment is a challenge in utility equipment planning. The practice of most utilities in this area so far is to use the deterministic method, which is basically based on an engineering judgment.

There are several reasons for the need for spares. First of all, a repairable failure of power equipment such as a transformer, reactor, capacitor, generator may often require a relatively long repair time. If the adequacy of equipment in a system is insufficient due to the lack of spares, the system may experience an extensive loss of energy supply and a financial loss of revenue. Second, equipment aging has been a major concern in utilities for years. Aged equipment implies a higher failure probability and thus a greater need for spares. Besides, the policy of common spares shared by an equipment group is becoming popular under the competitive environment in the power industry. Traditionally, for example, the $N - 1$ security principle has been widely used for substation transformers. Each substation is often designed to have two or more transformers in parallel so that the peak load can be still carried when one of the transformers fails. This is a secure but very expensive criterion. Compared to the $N - 1$ security principle in each substation, the common spare transformer strategy can avoid considerable capital expenditures and still assure a sufficient reliability level.

There are two types of spare-equipment schemes used in power systems. The off-line spare scheme refers to the spare in storage that will be installed to replace a piece of failed equipment. On-line redundancy is another spare-equipment scheme. For instance, some extra switchable reactors or capacitors are installed but not energized in the normal system state and could be switched in immediately when necessary.

Risk Assessment of Power Systems. By Wenyuan Li
ISBN 0-471-63168-X © 2005 the Institute of Electrical and Electronics Engineers, Inc.

The following are two basic questions in spare-equipment analysis:

1. How many spares are needed and when should each of them be in place in order to maintain system reliability?
2. How can spares be financially justified?

This chapter presents the probabilistic spare-equipment analysis techniques needed to answer these two questions. Section 11.2 discusses the spare-equipment analysis method based on reliability criteria and Section 11.3 addresses the spare-equipment analysis method using probabilistic cost models. The first method can be used separately or integrated into the second one as a unified procedure. Two examples are given in Sections 11.4 and 11.5. The first one is an analysis of off-line spare transformers and the second one is an analysis of on-line redundancy of 500 kV reactors.

11.2 SPARE-EQUIPMENT ANALYSIS BASED ON RELIABILITY CRITERIA [22, 30, 62]

Spares are considered for an equipment group. Each component in the group has its failure probability or unavailability and when it fails, a spare must be put in service to assure the normal operation of the system. Therefore, how many spares are needed depends on the requirement of group reliability. With the unavailability of individual components, Monte Carlo simulation or the state enumeration technique can be used to evaluate the group failure probability with and without spares. The spare-equipment analysis for an equipment group includes the following steps:

Step 1: Calculate the unavailability of components in the group
Step 2: Evaluate the individual failure event probabilities and the total group failure probability
Step 3: Perform the spare-equipment analysis based on a specified reliability criterion or a probabilistic cost model
Step 4: Repeat Steps 1 to 3 for all the years in consideration

11.2.1 Unavailability of Components

As discussed in Chapter 2, there are two failure modes for power system equipment: repairable failures and aging failures. In many risk evaluations, only repairable failures is considered. However, a model for the unavailability due to aging failures must be taken into consideration in the spare-equipment analysis since aging failure is one of the reasons why spares are needed, particularly for an aged equipment group.

11.2.1.1 Unavailability Due to Repairable Failures. As given in Section 2.2.1, the unavailability of a system component due to repairable failures is defined by

$$U_r = \frac{MTTR}{MTTF + MTTR} = \frac{f \times MTTR}{8760} \tag{11.1}$$

where $MTTR$ is the mean time to repair (hours), $MTTF$ the mean time to failure (hours), and f the average failure frequency (failures/year).

11.2.1.2 Unavailability Due to Aging Failures. The model for the unavailability due to aging failures has been developed in Section 2.2.2. For the system component with age T years, its unavailability due to the aging failure within the subsequent period t can be calculated using Equations (11.2) and (11.3):

$$U_a = \frac{1}{t} \sum_{i=1}^{N} P_i \cdot [t - (2i - 1)D/2] \tag{11.2}$$

where N is the number of the equal intervals into which the period t is divided, D is the length of each interval, and P_i is given by

$$P_i = \frac{\int_{T}^{T+iD} f(t)dt - \int_{T}^{T+(i-1)D} f(t)dt}{\int_{T}^{\infty} f(t)dt} \quad (i = 1, 2, \ldots, N) \tag{11.3}$$

where $f(t)$ is a two-parameter failure density probability function. If the normal distribution is assumed, the two parameters are the estimates of the mean life and standard deviation for the same type of equipment. If the Weibull distribution is used, they are the estimates of the scale and shape parameters. The analytical expressions of Equation (11.3) for the normal and Weibull distribution models have been given in Section 2.2.2.3.

11.2.1.3 Total Unavailability. The total unavailability of the two failure modes can be calculated using the union concept as follows:

$$U_t = U_r + U_a - U_r U_a \tag{11.4}$$

It is preferable to use the total unavailability if the state enumeration approach is used in the group risk evaluation. If the Monte Carlo simulation approach is used, two independent random numbers are created with one for U_r and another for U_a. The impacts of the unavailability due to both the repairable and aging failures can be automatically captured in the Monte Carlo simulation.

11.2.2 Group Reliability and Spare-Equipment Analysis

As mentioned above, evaluation of the group risk can be conducted using the Monte Carlo simulation or state enumeration technique. The procedure using the state enumeration method is given to explain the concept. Consider a three-component

Table 11.1 Event Probability

Event No.	Event	Event probability
1	1 down, 2 and 3 up	$U1 \cdot (1 - U2) \cdot (1 - U3)$
2	2 down, 1 and 3 up	$U2 \cdot (1 - U1) \cdot (1 - U3)$
3	3 down, 1 and 2 up	$U3 \cdot (1 - U1) \cdot (1 - U2)$
4	1 and 2 down, 3 up	$U1 \cdot U2 \cdot (1 - U3)$
5	1 and 3 down, 2 up	$U1 \cdot U3 \cdot (1 - U2)$
6	2 and 3 down, 1 up	$U2 \cdot U3 \cdot (1 - U1)$
7	all 1, 2, and 3 down	$U1 \cdot U2 \cdot U3$
8	all 1, 2, and 3 up	$(1 - U1) \cdot (1 - U2) \cdot (1 - U3)$

group. It is assumed that the unavailability values of the three components have been calculated and they are U1, U2, and U3. The event probability table is built as shown in Table 11.1. The cumulative failure probabilities for each failure level can be calculated from the table.

The probability for any one component failure is

$$P(a) = U1 \cdot (1 - U2) \cdot (1 - U3) + U2 \cdot (1 - U1) \cdot (1 - U3) + U3 \cdot (1 - U1) \cdot (1 - U2)$$

The probability for any two component failures is

$$P(b) = U1 \cdot U2 \cdot (1 - U3) + U1 \cdot U3 \cdot (1 - U2) + U2 \cdot U3 \cdot (1 - U1)$$

The probability for all the three component failures is

$$P(c) = U1 \cdot U2 \cdot U3$$

Given a system failure criterion, the spare-equipment analysis can be conducted. For instance, if the system failure criterion for this example is that any failure of one or more components results in the group failure, the spare-equipment analysis is shown in Table 11.2. Note that the reliability values in the "Example value" column are arbitrarily given here just for the purpose of explanation. If an acceptable group reliability level is specified, the number of spares can be determined. For instance, if the acceptable group reliability level is 0.9, the first spare is needed. If the acceptable level is selected as 0.98, the second one is also needed.

Table 11.2 Spare-equipment analysis based on a group reliability criterion

Spare	Group reliability	Example value	Spare contribution
Zero	$1.0 - [P(a) + P(b) + P(c)]$	0.85	
First	$1.0 - [P(b) + P(c)]$	0.95	0.10
Second	$1.0 - P(c)$	0.99	0.04
Third	1.0	1.00	0.01

11.3 SPARE-EQUIPMENT ANALYSIS USING THE PROBABILISTIC COST METHOD [22, 30, 62]

The spare-equipment analysis based on the group reliability requirement is useful and can be applied in some cases. However, specifying an acceptable group relia-bility level is a difficult task in actual applications. The probabilistic cost analysis provides a more comprehensive vehicle to justify or not justify spares. The analysis is based on the comparison between the group failure cost reduction due to spares and the investment cost of the spares.

11.3.1 Failure Cost Model

The failure cost is caused by component-failure events. The expected failure costs (EFC) for the cases without and with spares are calculated using the models in Equations (11.5) and (11.6), respectively:

$$EFC = \sum_{i=1}^{M} C_i P_i \cdot t_c \tag{11.5}$$

$$EFC = \sum_{i=1}^{M_1} C_i P_i \cdot t_c + \sum_{i=1}^{M_2} C_i P_i \cdot t_d \tag{11.6}$$

where C_i is the average failure cost per hour for the failure state i. P_i is the prob-ability of State i and can be obtained in Monte Carlo simulation or state enumer-ation. t_c is the total duration (in hours) in which the failure costs can take place. Obviously, t_c may be only a portion of the t in Equation (11.2). t_d is the installa-tion time of spares. M is the total number of failure states. M_1 is the number of the failure states in which the number of failed components exceeds the available spares. M_2 is the number of the failure states in which the failed components are equal or less than the available spares. In the failure states represented by M_2, the failure cost is due to the fact that installation of spares requires time. Normally, the contribution of the second term in Equation (11.6) is much smaller than that of the first term. In the case of on-line spares such as the switchable reactors or capacitors in transmission systems, the installation time is zero. In this case, only the first term is needed.

The failure costs for the cases of zero spare, one spare, two spares, and so on are calculated first. The differences in the failure cost between zero and one spare cases, one and two spare cases, and so on are the failure cost reductions due to the first spare, the second spare, and so on, which constitute the benefits in the bene-fit/cost analysis. It should be noted that according to the definition, M_1 and M_2 are varied for one, two, and so on spare cases. Generally, the calculation is repeated for each of all the years considered in spare planning. It should be also appreciated that P_i is different for each year because the unavailability of components due to the ag-ing-failure mode is increased with age.

11.3.2 Unit Failure Cost Estimation

The failure cost per hour C_i is an average value over the duration of t_c. The failure cost is generally caused by loss of loads in power systems. The three methods of estimating the unit interruption cost have been discussed in Section 1.2.3. A common situation in spare-equipment planning is that the loss of loads due to the lack of spares may be only associated with a reduction in the revenue to the utility. In this case, the failure cost per hour is equal to the loss of loads (in megawatts) times the electricity rate ($/megawatt hour). The loss of loads is time dependent, whereas the electricity rate may or may not be varied in the given time period. The electricity rate is basically a fixed price for the native customers within a utility and a time-varying value for the transactions between utilities.

In the case of fixed electricity rate, the average failure cost per hour is the product of the fixed electricity rate and the average loss of loads over the duration:

$$C_i = \frac{R}{N} \sum_{k=1}^{N} S_k \tag{11.7}$$

where R is the fixed electricity rate, S_k the loss of loads in the kth hour, and N the number of hours during t_c.

In the case of time-varying energy price, the average failure cost per hour can be calculated using the following approach. Both the loss of load S and energy price R are treated as random variables. Their experimental discrete probability distributions can be obtained through the hourly statistics of past years and expressed as a set of sample values, each having a probability:

$$\begin{aligned} p(S = s_j) &= p_{sj} \qquad (j = 1, \ldots, ns) \\ p(R = r_k) &= p_{rk} \qquad (k = 1, \ldots, nr) \end{aligned} \tag{11.8}$$

where s_j, p_{sj}, and ns are the sample values, probabilities of each value, and number of samples for the variable S; r_k, p_{rk}, and nr are the sample values, probabilities of each value, and number of samples for the variable R. In preparing the statistical curves of S and R, the hourly values can be divided into groups according to their magnitudes, using a given segmental length, and s_j or r_k as the average of the hourly values in a group. As long as the segmental length is small enough, the discrete probability distributions have sufficient accuracy.

The average failure cost per hour is given by

$$C_i = \sum_{j=1}^{ns} \sum_{k=1}^{nr} s_j r_k p_{sj} p_{rk} \tag{11.9}$$

Mathematically, the C_i in Equation (11.9) is the expectation of the product of the two random variables S and R, which is calculated as the sum of the products of each pair of s_j and r_k weighted by their probabilities of occurrence. Obviously, it includes the combinations of all the sample values of S and R.

11.3.3 Annual Investment Cost Model

Spares are associated with the capital investments that are translated into the cost in the benefit/cost analysis. The formula for the annual investment of a spare is the same as that used in Section 6.2.2:

$$A = V\frac{i(1+i)^n}{(1+i)^n - 1} \qquad (11.10)$$

where, A is the equivalent annual investment of a spare, V the actual capital investment in the year in which the spare is purchased, i the discount rate, and n the economic life of the spare.

In routine economic analysis, i is usually a fixed number obtained from the financial department of a utility. In fact, this number may vary from year to year. If there is information on the historical records of i in the past, the experimental probability distribution of i can be obtained. By applying the probability distribution assumption for i, the expected estimate of A can be calculated.

11.3.4 Present-Value Approach

The reductions in the failure costs due to spares and the equivalent annual investments for the spares are on a yearly basis and form two cash flows. The present-value approach is used to convert the cash flows into the present values and the benefit/cost analysis can be performed. The present value is calculated by

$$PV = \sum_{j=1}^{m} A_j/(1+i)^{j-1} \qquad (11.11)$$

where, PV is the present value, A_j the annual failure cost reduction or equivalent annual investment in year j, i the discount rate, and m the number of years considered in spare-equipment planning.

11.3.5 Procedure for Spare-Equipment Analysis

The procedure for spare-equipment analysis is summarized as follows:

- Calculate the unavailability of all the components in an equipment group for all the years considered.
- Conduct the spare-equipment analysis using the reliability criterion method presented in Section 11.2.
- Calculate the failure costs for zero spare, one spare, two spare, and so on cases using the models given in Section 11.3.1 and create a table of the failure cost reductions due to the spares for all the years in spare-equipment planning.
- Calculate the annual capital cost of a spare using Equation (11.10). If the capital cost is just less than the failure cost reduction due to a spare in a particular

year, one spare is considered for that year. The number and timing of the spares are marked on the failure cost reduction table.

- Calculate the two cash flows for the investment and failure cost reduction due to the spares.
- Calculate the present values of the two cash flows using Equation (11.11) and perform the benefit/cost analysis.

11.4   EXAMPLE 1: DETERMINING NUMBER AND TIMING OF SPARE TRANSFORMERS

11.4.1   Transformer Group and Data

This example is given to demonstrate the application of probabilistic spare analysis to the transformer group in which each transformer represents a single transformer substation. All the substations share common spare transformers. The analysis was performed in 2001. The group contained 16 transformers and their age ranged between 11 and 35 years, with an average of 27.8 years in 2001. The mean life of this type of transformer was estimated to be 42 years, with a standard deviation of 15 years. The repairable failure frequency and repair time of transformers were 0.072 failures/year and 155 hours/repair, respectively. This data was based on the historical statistics of the previous 15 years. Both aging and repairable failures were considered. A transformer failure would result in a loss of load and thus a failure cost. The age of each transformer and the average loss of load due to its failure are given in Table 11.3. The fixed energy price of $0.05/kWh was used to calculate the average failure cost per hour caused by loss of loads. Note that the average losses of loads in the table were for 2001 and an increase rate of 2% for each year was assumed in the study, based on the load growth forecast. The objective was to work out a spare plan for this group, indicating when and how many spares would be needed in the period from 2001 to 2020, and to conduct the benefit/cost analysis for financial justification. The investment cost of a spare transformer was estimated at $1.5 million. Its economic life was assumed to be the same as its mean life (42 years). In the application, a fixed discount rate of 8% was used.

11.4.2   Spare-Transformer Analysis Based on Group Failure Probability

The group failure probability is defined as the probability that some loads have to be curtailed due to any transformer failure. The enumeration technique described in Section 11.2.2 was used to perform the analysis. Table 11.4 shows the group failure probabilities with the different numbers of spare transformers for the 20 years. If the spare planning criterion is that the acceptable group failure probability should be smaller than 1%, it can be seen that we need the first two spare transformers in 2001, the third one in 2009, and the forth one in 2017.

Table 11.3 Transformer data

Transformer ID	Transformer age	Average loss of load (MW)
Aiy	13	7.2
Bab	33	3.6
Col	35	7.2
Dia	18	0.12
Fir	24	1.44
Fra	35	6.0
Gre	18	0.12
Isl	35	3.6
Lak	35	0.12
Mar	35	2.4
Mae	35	9.6
Mor	25	15.6
Por	23	2.4
Ste	11	3.6
Upp	35	7.2
70M	35	6.0

Table 11.4 Group failure probability

Year	Zero spares	One spare	Two spares	Three spares	Four spares
2001	0.18419	0.01658	0.00091	0.00003	0.0000009
2002	0.20420	0.02060	0.00127	0.00005	0.000002
2003	0.22580	0.02540	0.00175	0.00008	0.000003
2004	0.24879	0.03111	0.00240	0.00013	0.000005
2005	0.27329	0.03796	0.00327	0.00019	0.000008
2006	0.29922	0.04606	0.00442	0.00029	0.00001
2007	0.32650	0.05557	0.00591	0.00043	0.00002
2008	0.35504	0.06663	0.00786	0.00063	0.00004
2009	0.38471	0.07943	0.01035	0.00093	0.00006
2010	0.41540	0.09410	0.01353	0.00134	0.00010
2011	0.44692	0.11079	0.01754	0.00191	0.00015
2012	0.47911	0.12963	0.02254	0.00271	0.00024
2013	0.51176	0.15072	0.02873	0.00380	0.00036
2014	0.54467	0.17414	0.03630	0.00527	0.00055
2015	0.57759	0.19992	0.04548	0.00724	0.00084
2016	0.61031	0.22806	0.05649	0.00982	0.00125
2017	0.64258	0.25851	0.06956	0.01321	0.00183
2018	0.67417	0.29116	0.08492	0.01756	0.00266
2019	0.70486	0.32584	0.10278	0.02310	0.00382
2020	0.73443	0.36235	0.12333	0.03008	0.00541

11.4.3 Spare-Transformer Plans Based on the Probabilistic Cost Model

11.4.3.1 Evaluating Failure Cost Reductions Due to Spares. The failure costs due to the unavailability of transformers due to aging and repairable failures were evaluated using the models in Section 11.3. The differences between the cases without and with the spares provide the failure cost reductions that are shown in Table 11.5. The annual investment required for a spare transformer was estimated as $124.93 k using Equation (11.10). By comparing the annual investment for a spare transformer with the failure cost reduction due to a spare in different years, it can be seen that the first spare should be needed in 2001, the second one in 2007, and the third one in 2015. These are marked by * in Table 11.5. It is observed that the benefit provided by adding the next spare transformer, that is, the difference between the failure cost reduction and the annual investment, is always marginal at the beginning and increases over the years. This is because the unavailability due to the aging failures of transformers increases with the age.

11.4.3.2 Benefit/Cost Analysis. A spare-transformer plan has to be financially justifiable. The probabilistic cost method provides a vehicle to perform the benefit/cost analysis. Using the information in Table 11.5, the two cash flows for the investment cost and failure cost reduction associated with the three spares over the 20 years from 2001 to 2020 are obtained and shown in Table 11.6. The last line

Table 11.5 Failure cost reductions due to spares (k$/year)

Year	First	Second	Third	Fourth
2001	424.4*	38.2	2.1	0.08
2002	480.2	48.2	3.0	0.12
2003	541.3	60.8	4.3	0.19
2004	608.1	76.0	5.8	0.30
2005	680.6	94.4	8.2	0.48
2006	758.8	116.7	11.2	0.73
2007	842.8	143.3*	15.2	1.10
2008	932.5	174.9	20.6	1.66
2009	1027.7	212.0	27.6	2.46
2010	1128.3	255.3	36.7	3.62
2011	1233.7	305.5	48.4	5.27
2012	1343.8	363.2	63.1	7.58
2013	1458.1	429.0	81.7	10.79
2014	1575.8	503.3	104.8	15.18
2015	1696.3	586.6	133.3*	21.15
2016	1819.1	679.0	167.9	29.12
2017	1943.2	780.9	209.8	39.73
2018	2068.0	892.1	259.8	53.55
2019	2192.6	1012.4	318.8	71.40
2020	2316.3	1141.4	387.7	94.25

Table 11.6 Cash flows of investment costs and failure cost reductions for the spare plan (k$/yr)

Year	Investment	Failure cost reduction
2001	124.93	424.4
2002	124.93	480.2
2003	124.93	541.3
2004	124.93	608.1
2005	124.93	680.6
2006	124.93	758.8
2007	249.86	986.1
2008	249.86	1107.4
2009	249.86	1239.7
2010	240.86	1383.6
2011	249.86	1539.3
2012	249.86	1707.0
2013	249.86	1887.1
2014	249.86	2079.1
2015	374.79	2416.2
2016	374.79	2666.0
2017	374.79	2933.9
2018	374.79	3219.9
2019	374.79	3523.8
2020	374.79	3845.4
PV (2001)	2238.0	13583.5

indicates the present values (in 2001 dollars) of the two cash flows obtained using Equation (11.11). The benefit/cost ratio is 13583.5/2238 = 6.07. Apparently, the spare plan is investment effective.

This spare-transformer plan is different from the one based on the group failure probability criterion. If the four spares recommended in Section 11.4.2 are marked in the failure cost reduction table, a similar benefit/cost analysis can be conducted. The resulting benefit/cost ratio is 14147.1/3329.2 = 4.25. The spare-transformer plan based on the group failure probability criterion is less cost effective than the one based on the probabilistic cost method, although the former is still financially justifiable. On the other hand, if the three spares recommended in this section are marked in the group failure probability table, it can be found that the spare-transformer plan based on the group failure probability criterion has a higher reliability level than the one based on the probabilistic cost method. However, the latter is still acceptable from a reliability viewpoint. The analysis process provides the flexibility for a trade-off between the two spare-transformer plans.

Other similar benefit/cost analyses can be conducted. For instance, we can focus only on the first spare(s) for 2001 obtained using the probabilistic cost method or the group failure probability criterion. An actual financial investment decision often emphasizes the immediate need. In this case, only the years prior to the next spare

Table 11.7 Benefit/cost ratios for the four spare-transformer plans

Plan	Benefit/cost ratio
Plan 1	6.07
Plan 2	4.25
Plan 3	4.54
Plan 4	2.87

are considered, that is, the first 6 years in the cash flow for the probabilistic cost method and the first 8 years for the group failure probability criterion. The benefit/cost ratios for the two long-term and two short-term spare-transformer plans are summarized in Table 11.7. These spare plans are:

Plan 1: from 2001 to 2020; the first spare in 2001, the second one in 2007, and the third one in 2015 (using the probabilistic cost method)

Plan 2: from 2001 to 2020; the first two spares in 2001, the third one in 2009, and the fourth one in 2017 (using the group failure probability criterion)

Plan 3: from 2001 to 2006; one spare in 2001 (using the probabilistic cost method)

Plan 4: from 2001 to 2008; two spares in 2001 (using the group failure probability criterion)

The results for the two short-term spare-transformer plans indicate that one immediate spare in 2001 is financially justifiable to assure the reliability. If a higher reliability level is preferred, two spares in 2001 are still acceptable according to financial justification.

11.4.4 Summary

The two long-term and two short-term spare-transformer plans are worked out using the probabilistic spare analysis. All the four plans are cost-effective and meet the reliability requirement of the transformer group.

Two plans are obtained using the group failure probability criterion and the other two using the probabilistic cost method. The group failure probability criterion alone may be confined in some actual applications since specifying an acceptable failure probability usually requires a difficult engineering judgment. The probabilistic cost method offers the quantified benefit/cost analysis that can lead to a direct financial justification.

11.5 EXAMPLE 2—DETERMINING REDUNDANCY LEVEL OF 500 kV REACTORS [22, 74]

This example describes the application of the probabilistic spare analysis method in determining the redundancy level of 500 kV reactors.

11.5.1 Problem Description

High-voltage reactors are installed in long-distance transmission systems to protect system equipment from damage caused by transient or steady-state overvoltage in a sudden loss of a tie line with high export power. The numbers and locations of reactors are determined in the total transfer capability (TTC) calculations through considerable pre- and postoutage power flows and transient and voltage stability studies in the normal and various contingence system states. Historically, the redundancy of reactors was not a major concern. This was because, first of all, the failure probability of high-voltage reactors was low. Second, even if a failure of reactors took place, it would not be critical under the previous energy trade environment. The energy exchanges on tie lines were generally based on a long-term contract. If a reactor outage resulted in the temporary reduction of energy sales due to the decreased export transfer limit, it was relatively easy for a purchaser to buy the backup energy at almost the same price. The seller would make up the due energy later and still meet the long-term contract.

A fundamental change in the power market has occurred since the power industry entered the deregulation era. Not only have the energy trade activities on tie lines become much more frequent but also the average power trade level is much higher today. Particularly, the real-time energy-trade model has become the norm and electricity rates are extremely volatile and heavily time dependent. If a 500 kV reactor outage causes a reduction in export capability even for a short duration, it may be associated with a huge loss of potential income or a high financial penalty. Therefore, the redundancy of reactors becomes an important consideration. Although the redundancy may not increase the maximum TTC of a transmission system, it enhances the profile of the dynamic TTC. The questions faced by a utility are: Is the redundancy really needed in the existing system or is it worth providing? If yes, how much redundancy of 500 kV reactors can be economically justified? On the one hand, the cumulative time during which the export can reach the maximum transfer capability on the tie lines is a small portion of one year. This implies that the probability of a reactor failure event occurring within this period is low. On the other hand, transmission equipment, including reactors, is gradually aging. The aged reactors increase the failure probability.

Obviously, this problem can be resolved using the probabilistic spare-equipment analysis method presented in Section 11.3, with the following features:

- In the system control center of a utility, operation planning engineers prepare many operation orders. One of them is the operational requirement for the reactors associated with the system TTC, which generally includes a statement similar to the following: All the 500 kV shunt reactors must be available to assure the TTC of X MW on the tie lines. Otherwise, any one or more reactors out of service results in the reduction of the TTC to Y MW from X MW. Whether or not a reactor outage event causes an economic loss depends on whether the power needed by energy buyers is higher or lower than Y MW during the outage. It can be seen that the economic loss due to a reactor outage event is caused by a possible loss of megawatt export, which can be determined by

$$s_j = \max\{0, T_j - Y\} \tag{11.12}$$

where s_j is the hourly loss of megawatt export, which is similar to the loss of load defined in Equations (11.8) and (11.9); T_j the power needed by energy buyers, which is the flow on the tie line in the jth hour; and Y the decreased TTC limit during a reactor outage event.

- Although the Y is a deterministic value in the operation order, the export power flow T_j on the tie line is time dependent so that the s_j is randomly varied. The energy price in this case entirely depends on the power market. In other words, both the real-time energy price and loss of export power are the two random variables.
- The t_c in Equations (11.5) and (11.6) is only the small portion of one year during which the export demand may be higher than the transfer capability Y, since high-demand transactions generally take place merely within those several months of a year.

The concept of performing the redundancy analysis of 500 kV reactors is similar to that for the spare-equipment analysis. The expected economic loss due to the unavailability of 500 kV reactors can be evaluated for different redundancy levels. The differences in the expected economic loss between the redundancy levels reflect the economic worth or benefit provided by the redundant reactors. The benefits due to the redundant reactors can be compared with their capital investments.

11.5.2 Study Condition and Data

This is an actual application in a utility system. There were 47 sets of 500 kV reactors in the system. Both the repairable- and aging-failure modes were modeled. The repairable-failure frequencies and repair times of the reactors were obtained from the statistical records of the previous 15 years. The age of the reactors ranged from 8 to 32 years, with half of them beyond 30 years. In the aging-failure modeling, a mean life of 37.6 years with a standard deviation of 6.9 years was used. These data were obtained using the aging-failure statistics of reactors and the parameter estimation method of Section 3.7.1. Table 11.8 shows the unavailability values of the reactors in 2001 for both the repairable- and aging-failure modes. Note that the unavailability values due to the aging failure would be increased over the years beyond 2001.

Figure 11.1 is a statistical curve of the actual export powers on the tie line flowing from the utility to its neighbor, which is based on the data of a few months. These months correspond to the high-transaction period during which the exported power may exceed the limit Y. The line with an arrow is 0 MW, so that the part of the curve above the line corresponds to the export and the part below to the import. The other bold line is the megawatt level of the limit Y in the operation order. Fig-

Table 11.8 Unavailability of 500 kV reactors in 2001

No.	Repairable	Aging	No.	Repairable	Aging
1	0.00283	0.00001	25	0.00412	0.01928
2	0.00930	0.00001	26	0.00027	0.00162
3	0.00157	0.00000	27	0.00098	0.00162
4	0.00040	0.00001	28	0.00039	0.00162
5	0.00016	0.00001	29	0.00856	0.00162
6	0.00040	0.00001	30	0.00013	0.00162
7	0.00040	0.00001	31	0.00076	0.00162
8	0.00075	0.00001	32	0.00064	0.00162
9	0.00040	0.00001	33	0.00209	0.01928
10	0.00040	0.00001	34	0.00033	0.00007
11	0.00040	0.00001	35	0.00309	0.00001
12	0.00012	0.00000	36	0.00309	0.00001
13	0.00004	0.00001	37	0.00309	0.00001
14	0.00019	0.00001	38	0.00309	0.00001
15	0.00012	0.00000	39	0.00309	0.00001
16	0.00015	0.00027	40	0.00309	0.00001
17	0.00005	0.00007	41	0.00224	0.00051
18	0.00003	0.00000	42	0.00247	0.02577
19	0.00199	0.00027	43	0.00247	0.02577
20	0.00262	0.02577	44	0.00247	0.02577
21	0.00262	0.02577	45	0.00247	0.02577
22	0.00379	0.01928	46	0.00247	0.02577
23	0.00379	0.01928	47	0.00247	0.02577
24	0.00346	0.01928			

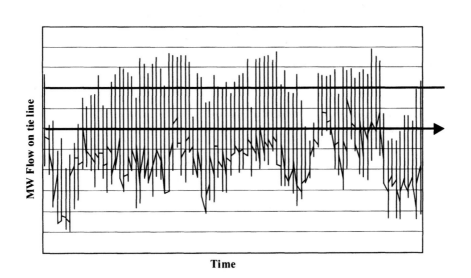

Figure 11.1 Statistical curve of the export power flow on the tie line.

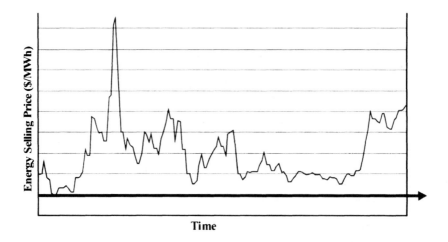

Figure 11.2 Statistical curve of the real-time energy price.

ure 11.2 shows a statistical curve of the real-time energy price at the border between the two utilities during the same months. The line with an arrow denotes the level of 0 $/MWh. The discrete probability distributions of the export loss and energy price can be obtained from the two curves.

11.5.3 Redundancy Analysis

The expected economic losses due to the unavailability of 500 kV reactors at the different redundancy levels for 2001 to 2006 were evaluated using the probabilistic cost method and are given in Table 11.9. With the information in the table, the incremental economic benefits created by the first to the fourth redundant reactors, which are the reductions in the expected economic loss, can be calculated and are shown in Table 11.10. The following observations can be drawn:

Table 11.9 Expected economic losses due to unavailability of reactors at different redundancy levels (M$/yr)

	Redundancy level				
Year	Zero reactors	One reactors	Two reactors	Three reactors	Four reactors
2001	6.68	1.21	0.15	0.016	0.0005
2002	8.00	1.79	0.27	0.03	0.005
2003	9.40	2.56	0.45	0.05	0.010
2004	10.88	3.58	0.80	0.14	0.025
2005	12.30	4.78	1.27	0.25	0.035
2006	13.74	6.21	1.99	0.49	0.120

Table 11.10 Incremental economic benefits due to redundant reactors (M$/yr)

Year	First reactor	Second reactor	Third reactor	Fourth reactor
2001	5.47*	1.06*	0.134	0.0155
2002	6.21	1.52	0.24	0.025
2003	6.84	2.11	0.40	0.040
2004	7.30	2.78	0.66*	0.115
2005	7.52	3.51	1.02	0.215
2006	7.53	4.22	1.50	0.370

- The first redundant reactor provides much more economic benefits than the others. This is due to the fact that the probability of first-level failure events (one reactor failure) is much larger than that of higher-level failure events (failures of more than one reactor).

- The economic benefits due to the redundant reactors increase over the years. This is because the aging-failure probability of the reactors will increase with the age. In this application, the export power profile was assumed to be identical from 2001 to 2006.

In order to identify a rational redundancy level, benefit/cost analysis should be performed. A redundant reactor requires investment capital and provides economic benefits. A three-phase 500 kV reactor is estimated to cost $6.5 million. By applying the economic life of 38 years (i.e., the mean life of 500 kV reactors) and the discount rate of 8%, the annual investment cost of a reactor is $0.55 million. If the annual cost is smaller than the annual benefit for a particular year in Table 11.10, a redundant reactor can be financially justified. Using this comparison, we can see that the first two redundant reactors should be added in 2001, followed by the third one in 2004. They are marked by * in Table 11.10.

If only the first two redundant reactors for the year 2001 are considered, the benefit/cost ratio is $(5.47 + 1.06)/(0.55 \times 2) = 5.94$. If the six year period from 2001 to 2006 is considered, the present-value approach is applied to convert both the costs and benefits for each year in this period to the first year to include the time value factor of money. The two "cash flows" for the annual investment costs and economic benefits with their present values (PV) in 2001 dollars are shown in Table 11.11. The benefit/cost ratio for the 6 year period is $47.76/6.71 = 7.12$. Apparently, the high benefit/cost ratio either for the year 2001 or for the 6 year period leads to confidence in the financial justification of the reactor redundancy.

11.5.4 Summary

The basic idea presented in the example is to determine the rational redundancy level of 500 kV reactors based on the coordination among the energy price in the pow-

Table 11.11 Cash flows of the annual investment costs and economic benefits (M\$/yr)

Year	Investment cost	Economic benefit
2001	1.10	6.53
2002	1.10	7.73
2003	1.10	8.95
2004	1.65	10.74
2005	1.65	12.05
2006	1.65	13.25
PV (2001)	6.71	47.76

er market, the total transfer capability (TTC) in the operation order of the utility control center, and random equipment failures.

In the economic aspect, the redundancy level of 500 kV reactors determined using the presented method enables one to avoid the possible losses in the income from energy sales. In the technical aspect, the redundancy level improves the profile of the system TTC in such a way that the maximum export transfer capability can be applied. This reduces the complexity of using variable transfer capabilities in operation. Particularly, the redundancy level obtained guarantees a high benefit/cost ratio.

Although the example is associated with the redundancy of 500 kV reactors, the concept and the approach are general and can be applied to other on-line reactive equipment.

11.6 CONCLUSIONS

This chapter discusses the methods of probabilistic spare-equipment analysis and their applications. The basic concept of spare-equipment analysis is straightforward. A spare can reduce the failure probability of an equipment group and thus the failure cost, resulting in economic benefits but it also requires capital investment. Two methods have been discussed. The first one is based on the reliability criteria of an equipment group and the second one on the probabilistic cost model. The second method is more comprehensive since the benefit/cost analysis is conducted, which provides the financial justification for decision making.

The key to the probabilistic cost method is the failure cost model. The three factors in failure cost evaluation are the probability of system failure states, the loss of loads or power exports, and the energy price. In general, all the three are random variables, although the energy price may be a fixed rate in some cases. The probability of system failure states depends on the unavailability of system components and can be evaluated using either Monte Carlo simulation or the state enumeration technique. It is worth stressing that not only the unavailability of system components due to repairable failures but also the unavailability due to aging failures should be included in the spare-equipment analysis, particularly for a long-term spare-equipment plan. Randomness of loss of load or energy price is represented using a discrete probability distribution that can be obtained from historical data records. The expectation of the product of loss of load and energy price is the estimate of the average hourly failure cost over the time period.

Two actual examples are given to demonstrate the application. The first one is associated with the spare-equipment planning for transformers. The second one addresses the redundancy issue of on-line reactive equipment using 500 kV reactors as an example. This is an interesting topic under the deregulated power-market environment. The basic method for the redundancy analysis of on-line equipment is similar to that for the off-line spare-equipment analysis. A crucial step is to clarify the relationship between the spare-equipment analysis model and some special concepts in power system operation, such as the operation order in utility control centers, total transfer capability (TTC), and real-time energy price.

An important point to appreciate is that the spare-equipment analysis should be updated every year for a system or an equipment group. This is not only because the aging-failure unavailability of components will increase over the years but also because other data like the load forecast and energy price may be changed.

CHAPTER 12

RELIABILITY-BASED TRANSMISSION-SERVICE PRICING

12.1 INTRODUCTION

The electric power utilities around the world have been undergoing rapid changes in business environment in the past few years. One of the most prevalent changes is unbundling of services, leading to the open access to transmission. Many utilities have opened their transmission systems to different power suppliers and consumers. In the new structure of the power industry, the function of transmission becomes a service to all the types of customers including loads, generators, and wheelers. The transmission-service pricing method must be designed to respect the following principles:

- Fairness: treating all the customers equally
- Economic efficiency: allowing the full recovery of investment requirements and encouraging the right siting of new generators
- Reliability: forcing all the players in the system to share the responsibility for system reliability

In general, transmission-service pricing involves the following components:

- Value of existing transmission system use
- Expenditure on operation, maintenance, and administration
- Investment in system reinforcement
- Capital for replacement and refurbishment of old equipment
- Ancillary service cost
- System reliability cost
- Opportunity cost

Risk Assessment of Power Systems. By Wenyuan Li
ISBN 0-471-63168-X © 2005 the Institute of Electrical and Electronics Engineers, Inc.

The available methodologies for transmission pricing can fall into three categories: (1) embedded cost methods, (2) marginal or incremental cost methods, and (3) market-driven methods (power auction models). Most of the methodologies do not include all the components mentioned above. Particularly, the investigation into the system reliability cost component is very limited.

System reliability should be considered as a component in transmission pricing. This is because any access of a new wheeler or load to the system causes an incremental increase in system risk (unreliability). The incremental risk is the potential damage cost and represents the potential requirement for transmission reinforcement. Reliable system operation must be closely tied to the price signals that induce participants to act in the ways that share the responsibility for overall system reliability.

This chapter discusses a method to calculate the reliability component in transmission pricing. The reliability component must reflect the long-run incremental investment requirement; treat the native, wheeling, and generation customers equally based on their impacts on system reliability; and send a price signal to encourage the right siting of generators (such as independent power producers). The basic concept, calculation method, and rate design are discussed, respectively, in Sections 12.2, 12.3, and 12.4. An application example is given in Section 12.5.

12.2 BASIC CONCEPT [67, 77]

There are three tasks required to include the reliability component in transmission pricing. The first task is to quantify the risk cost or reliability value. The three methods of assessing the unit interruption cost have been presented in Section 1.2.3. The first one is based on the customer damage functions, the second one on the capital investments, and the third one on the electricity elasticity coefficient. The first and third methods are extensively used in the applications associated with system planning and operation, which have been discussed in the previous chapters. These two methods, however, are not applicable to transmission pricing since the customer damages or economic losses reflected in GDP (gross domestic product) due to system risk are not the costs related to transmission services. The second method is suitable for transmission pricing and is described in detail here. The second task is to quantify the impacts of the customers on transmission-system risk. The composite generation and transmission risk-evaluation method provides an appropriate tool for this purpose. The third task is to incorporate the reliability component into the rate design. One important principle is that the rate, including the reliability component, must be simple and easy to implement.

12.2.1 Incremental Reliability Value

From an overall system viewpoint, the amount of load growth requires the same amount of increase in the transmission-system transfer capability to maintain the constant system reliability level. In other words, one megawatt of load growth re-

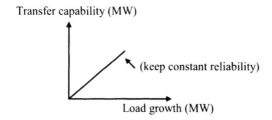

Figure 12.1 Correlation between load growth and transfer capability requirement.

quires a one megawatt increase in the system transfer capability in the sense of long-term and system average if we want to keep the system reliability unchanged. This concept is shown in Figure 12.1. According to the concept of long-run incremental cost, the capital investment for system reinforcements can be calculated and expressed as the unit investment cost (UIC) which is dollar-per-year for each megawatt increase in the system transfer capability. Utilities calculate the UIC through a series of system-reinforcement project-planning studies. The relationship between transfer capability and capital investment is shown in Figure 12.2. By combining this second concept with the first one, the UIC is equivalent to dollars per year of the capital expenditure required for each megawatt of system load growth to maintain the constant system reliability level.

As the load grows, the system risk or unreliability increases as shown in Figure 12.3. Generally, such an increase curve is nonlinear. It is assumed that when the load growth reaches some point (such as Point B in the figure), a transmission reinforcement project (such as a line addition) is put in service and brings the system risk back to the initial level. This corresponds to the capital investment of the project required to cover the total load growth. In other words, the increased system risk due to *B* MW load growth is just offset by the system reinforcement required by the same *B* MW load growth. The system still remains at the same risk level as it was at the beginning. Before the reinforcement is physically in place, the system reliability gradually deteriorates due to the load growth. However, the deterioration is

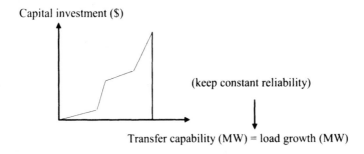

Figure 12.2 Relationship between transfer capability and capital investment.

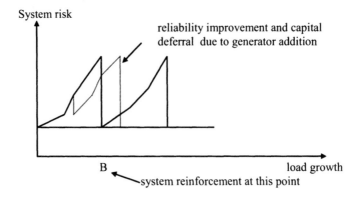

Figure 12.3   Correlation between system risk and load growth.

thought to be in an acceptable range. As a matter of fact, the risk in any system al-ways fluctuates up and down in the long run. It goes up with the load growth and then comes down until a system reinforcement measure is completed. Based on this analysis, the average unit incremental unreliability (AUIU) can be defined as the average variation in a risk index due to each megawatt of load growth. Obviously, the AUIU can be calculated using a system risk evaluation method. Conceptually, any risk index can be used because only the relative variation in the index is needed. It is suggested to use the expected energy not supplied (EENS) since this index varies continuously with load growth.

The unit incremental reliability value (UIRV) can be obtained by dividing the UIC by the AUIU. The UIRV is an indicator of reliability value (\$/year/unit relia-bility variation) and can be used in designing a charge or credit rate for different customers. If a customer causes deterioration of system reliability, it should be charged. Otherwise, if it improves system reliability, it could be credited.

12.2.2   Impacts of Customers on System Reliability

The customers that use transmission services can be categorized into three types. The first type is called native or network load customers. They are "pure" loads and need both generation and transmission services. The second type is classed as gen-erator customers. They are "pure" generations and sell powers to the utility through the transmission access. The third type is called wheeling customers. They need the transmission access to transport electric power to their own consumers. If a load is added to the system, it generally causes an incremental increase in transmission sys-tem risk and therefore it should be charged by the rate of the UIRV based on its negative impact on system reliability. If a generator is added to the system, it gener-ally causes an incremental decrease in transmission system risk, particularly when it is located near the load center and, therefore, it should be credited by the same rate (negative charge). A wheeling customer provides a load addition and the same amount of generation addition but at different locations. As a result, the impacts of its load and generation additions need to be calculated separately and valued with the same rate. It should be noted that for simplicity, the same amount of load and generation is attributed to the wheeling customer in the discussion. The loss due to a wheeler can be easily considered using different methods. For instance, the loss is often considered as an ancillary service using a separate charge rate. The incremental variations in transmission system risk due to a generator or a load at different locations are different and can be evaluated using the composite generation and transmission risk evaluation method. In the evaluation, all generation sources are assumed to be 100% reliable and only the outages and capacity limits of transmission components are considered so that the evaluated impacts are those on the transmission system risk.

It should be noted that although the UIRV is called as the "reliability value," it is deduced from the capital investment required by load growths and therefore fully corresponds to the revenue requirement for system reinforcements. Introduction of the reliability index provides a vehicle to differentiate the impacts of individual generations and loads and quantify their values. It can be seen that the UIRV is the charge or credit rate fair to all the types of customers. Such a reliability component offers the credit to a generator (including the equivalent generation of wheeling customers) and the amount of the credit relies on the impact, which is dependent on its location. As a result, it becomes an incentive factor to encourage the right siting of new generators. A generator at a right site reduces the system risk and plays a role in deferring the transmission reinforcement. This concept is also shown in Figure 12.3.

12.2.3 Reliability Component in Price Design

The charge for loads or the credit for generations calculated using the above concepts is an additional component on top of the base charge rate in the transmission service price design. It reflects negative or positive contributions of new loads or new generators to the system reliability, which is positively correlated to system investments. The price design is not a pure technical issue. Different utilities may chose different pricing calculation methods based on their market models and business considerations. Since the reliability component developed is the additional portion independent of the base rate, this provides flexibility in the application. The base rate can be calculated using any conventional method including nontechnical considerations.

The reliability component is a charge term for a load customer and a credit term for a generator customer. It includes both the terms for a wheeling customer. Besides this, there should be another term for each load, including native and wheeling customers. This additional term corresponds to the benefit obtained by a load from the reliability improvement produced by generation customers. In the rate design, the credits assigned to the generators (including the equivalent generation of the wheeling customers) should be commonly borne by the utility and all the load customers, including both the native and wheeling customer's loads. This is because, on the one hand, every load gets the benefit from the system reliability improvement due to generation additions and, on the other hand, the utility also acquires the benefit from the potential deferral of system reinforcement due to the same generation additions. In other words, only one portion of the credits to the generators is born by the load customers and the rest should be absorbed by the utility or transmission company itself.

It can be seen that the reliability component in the price design provides an incentive signal to both the utility and its customers to share the responsibility for overall system reliability.

12.3 CALCULATION METHODS [67, 77]

With the concepts described above, the following calculation methods are developed.

12.3.1 Unit Incremental Reliability Value (UIRV)

The calculation of the UIRV follows these three steps:

1. The unit investment cost (UIC) is calculated based on the transmission expansion planning studies for a series of reinforcement projects that enhance the system transfer capability in response to the load growth. The unit is in $/year/MW.

2. The year in which a system reinforcement project was completed is selected as a reference. For the period between the reference year and the year just before the next reinforcement project was completed, two transmission system risk evaluation studies are performed using the system configuration before the second reinforcement. The first study uses the load level at the reference year and the second study uses the load level at the end year of the period. The process is repeated for as many as possible reinforcement project periods. The average unit incremental unreliability (AUIU) is calculated by

$$AUIU = \frac{1}{N} \sum_{i=1}^{N} \frac{EENS_{ei} - EENS_{si}}{LG_i} \tag{12.1}$$

where $EENS_{si}$ and $EENS_{ei}$ are the risk indices of expected energy not supplied for the start and end years in the ith reinforcement period, LG_i is the system load growth between the start and end years in the period and N is the number of the reinforcement periods considered.

3. The unit incremental reliability value (UIRV) is calculated by

$$UIRV = \frac{UIC}{AUIU} \tag{12.2}$$

The unit is $/year/unit incremental EENS.

12.3.2 Generation Credit for Reliability Improvement (GCRI)

For a generator (a real generator or equivalent generator of a wheeling customer), the two cases for the existing system are studied using the transmission risk evaluation method: one case without the generator and another with it. In the evaluations, only transmission component failures are considered. The difference between the EENS indices for the two cases represents the improvement in system reliability due to the new generator. The generation credit for reliability improvement (GCRI) is calculated by

$$GCRI_i = (EENS_{Bi} - EENS_{Gi}) \times UIRV \times AF_i \qquad (12.3)$$

where the $EENS_{Bi}$ and $EENS_{Gi}$ are the EENS indices for the cases without and with the ith generator, respectively. AF_i is the availability factor of the generator. A generator cannot be 100% available. When the generator is unavailable, its positive impact on system reliability will no longer exist. For an equivalent generation of a wheeling customer, AF_i is specified in the service agreement. For a real generator, AF_i is estimated by

$$AF_i = 1.0 - (FOP_i + MOP_i + POP_i + OOP_i) \qquad (12.4)$$

where FOP_i is the forced outage probability, MOP_i the maintenance outage probability, POP_i the planned outage probability for nonmaintenance reasons, and OOP_i the other outage probability.

12.3.3 Load Charge for Reliability Degradation (LCRD)

For a load (a new native or wheeling customer's load), the two cases for the existing system are studied using the transmission risk evaluation method: one case without the load and another with it. The difference between the EENS indices for the two cases represents the degradation in system reliability due to the load. The load charge for reliability degradation (LCRD) is calculated by

$$LCRD_i = (EENS_{Li} - EENS_{Bi}) \times UIRV \qquad (12.5)$$

where $EENS_{Bi}$ and $EENS_{Li}$ are the EENS indices for the cases without and with the ith load. The $EENS_{Bi}$ is the same as that in Equation (12.3).

It should be noted that no factor like the AF_i is introduced in calculating the load charge for reliability degradation. This is because of the fact that the amount of the system reinforcement required by a new load to keep system reliability unchanged depends on its peak value and is not reduced even if the load needs to be served only for the partial time of one year.

12.3.4 Load Charge Rate Due to Generation Credit (LCRGC)

As mentioned earlier, the total credit assigned to the generation customers (real generators and equivalent generators of wheelers) is commonly borne by the utility or transmission company and all the load customers (native and wheeling loads). The LCRGC is calculated by

$$LCRGC = \frac{SF \times \sum_{i=1}^{M} GCRI_i}{EL + IL + \sum_{j=1}^{L} WL_j} \qquad (12.6)$$

where $GCRI_i$ is the generation credit given to the ith generator, WL_j is the peak load of the jth wheeling customer, EL the total existing native peak load, IL the increased native peak load, M the total number of real generators and equivalent generators, L the number of wheeling customers, and SF the share factor between the utility and load customers, which is generally 0.5.

12.4 RATE DESIGN

12.4.1 Charge Rate for Wheeling Customers

A wheeling customer includes both generation and load and its charge rate is

$$CR_j = BR_j + \frac{LCRD_j}{WL_j} - \frac{GCRI_j}{WL_j} + LCRGC \qquad (12.7)$$

where the subscript j indicates the jth wheeling customer, CR_j is the composite charge rate, BR_j the base rate before incorporating the reliability component, WL_j the peak wheeling load, $LCRD_j$ the load charge for reliability degradation for the wheeling customer, $GCRI_j$ the generation credit for reliability improvement for the wheeling customer, and $LCRGC$ the additional load charge rate due to generation credit.

The first term is the base rate. The second term corresponds to the wheeling load charge, and the third term to the wheeling generation credit (negative charge). The fourth term, $LCRGC$, reflects the additional charge due to the benefit obtained by the wheeling customer from the system reliability improvement caused by all generation additions, including its own generation.

12.4.2 Charge Rate for Native Customers

In the case of native customers, only new loads cause reliability deterioration. There are two rate design principles for the native customers. The first one is the overall system postage stamp method—no discrimination is made between old and new loads. Based on this method, the load charge for the reliability degradation caused by the new loads is distributed in the total system loads, leading to the same adjustment to the base rate for all the native customers. The second one is the localized postage stamp method—the increased load charge only applies to the new loads. Equation (12.8) is utilized for the first principle. In this case, all the native customers have the identical rate:

$$CR = BR + \frac{\sum_{i=1}^{MI} LCRD_i}{EL + IL} + LCRGC \tag{12.8}$$

where CR is the composite charge rate for the native customers, BR the base rate for the native customers, $LCRD_i$ the load charge for reliability degradation for the ith new native load, $LCRGC$ the additional load charge rate due to generation credit, EL the total existing native peak load, IL the increased native peak load, and MI the number of new native loads.

If the second principle is applied, the denominator of the second term in Equation (12.8) becomes IL only for the new loads and the whole second term does not exist for the old customers without any load growth.

12.4.3 Credit to Generation Customers

For a generation customer, the annual credit is the GCRI. There is no need to calculate the rate per megawatt. It is important to recognize that the credit is only due to the contribution to the transmission system reliability improvement and should not be confused with the income of selling the electric energy produced by the generator.

12.5 APPLICATION EXAMPLE [70, 77]

A utility system is used as an example to demonstrate the calculation procedure of the presented method. This example is only for the research purposes and the result was not used in an actual application. The utility owns generation and transmission facilities. Under the environment of deregulated power industry, the utility provides open access to transmission for external power suppliers and wheeling customers and must treat them the same way as its own generators.

The utility system includes the four main service regions that are expressed by the abbreviations PR, SI, LM, and VI. The PR and SI regions are the generation centers and the LM and VI regions are the load centers with a small number of generation sources. The skeleton of the transmission service regions and customer locations is shown in Figure 12.4. The symbols of G and L in the figure denote the generation and load points, respectively. The transmission risk evaluation in the study was conducted on the 500 kV and 230 kV systems that cover all the four service regions. The customer information is as follows:

- The existing system peak load in the study case was 7537 MW
- There was 2% load growth for each native customer as new loads

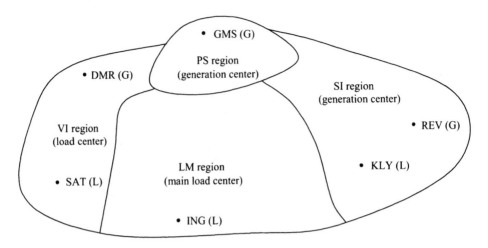

Figure 12.4 Skeleton of transmission service regions and customer locations.

- There were three wheeling customers:
    50 MW from Bus DMR in the VI region to Bus KLY in the SI region
    50 MW from Bus GMS in the PR region to Bus SAT in the VI region
    50 MW from Bus REV in the SI region to Bus ING in the LM region
- There was one cogeneration customer: 50 MW at Bus DMR in the VI region

12.5.1 Calculation of the UIRV

The average value of the unit investment cost (UIC) is $36k/year/MW. This is based on the system expansion planning studies for a series of system reinforcement projects. The average unit incremental unreliability (AUIU) is 7.539 MWh/year/MW. This number is obtained using Equation (12.1) and the same reinforcement projects as in calculating the UIC. A risk evaluation program of composite generation and transmission systems is used to assess the system EENS indices required in Equation (12.1) for the start and end years in each reinforcement period. The UIRV is calculated using Equation (12.2):

$$UIRV = \frac{\$36 \text{ k}}{7.539} = \$4775/\text{year/unit } EENS$$

Note that the unit of EENS is MWh/yr.

12.5.2 Calculation of the GCRI

The three wheeling and one cogeneration customers are associated with the three generator sources: 100 MW at Bus DMR in the VI region, 50 MW at Bus GMS in

Table 12.1   EENS indices and GCRI

New generator	EENS (MWh/yr)	GCRI (k$/yr)
Base system	3580	
100MW at DMR (VI)	2861	2918
50 MW at GMS (PR)	3502	317
50 MW at REV (SI)	3540	162

the PR region, and 50 MW at Bus REV in the SI region. The same risk evaluation program is utilized to assess the EENS indices for the existing system and the system cases with each of these three generators. Equation (12.3) is employed to calculate the GCRI, and the availability factor (AF) of the generators is assumed 85%. The results are shown in Table 12.1. It can be seen that the generation credits for the generators in the different regions are varied. The generator in the VI region will obtain much more credit than the generator located in the PR or SI region since the VI region is a load center, whereas the PR and SI regions are generation centers. This provides a pricing signal of encouraging new generations in the VI region.

12.5.3 Calculation of the LCRD

The three wheeling customers are associated with the three 50 MW loads at Bus SAT in the VI region, Bus ING in the LM region, and Bus KLY in the SI region, respectively. The risk evaluations are performed to calculate the EENS indices for the existing system, the system cases with each of these three wheeling loads, and the system case with the 2% native load growth. Equation (12.5) is utilized to calculate the LCRD. The results are shown in Table 12.2.

It can be seen that the load charges for the wheeling loads located in the different regions are also varied. The load in the VI region, which is far away from the generation centers, will be charged more than the load located in the LM region, which is relatively near the generation centers. It will be charged much more than the load located in the SI region, which is a generation center. The 2% load growth of each native customer corresponds to the total increase of 151 MW. It can be seen that the

Table 12.2   EENS indices and LCRD

New load	EENS (MWh/yr)	LCRD (k$/yr)
Base system	3580	
50 MW at SAT (VI)	3876	1413
50 MW at ING (LM)	3805	1074
50 MW at KLY (SI)	3725	692
2% load increase (at each native load)	4335	3605

load charge for the 151 MW of new loads uniformly distributed in the system is more than the total load charge for the 150 MW loads of the three wheeling customers.

12.5.4 Calculation of the LCRGC

There are three new generators and the total credit assigned to them is 2918 + 317 + 162 = 3397 k$/year. The existing native peak load is EL = 7537 MW, the increased native peak load is IL = 7537 × 2% = 151 MW and the three wheeling loads are WL = 50 × 3 = 150 MW. The share factor is assumed to be 0.5. The LCRGC is obtained using Equation (12.6) and is $0.217/kW/year. The LCRGC is always a very small value.

12.5.5 Calculations of Charge Rates

The base rate is assumed to be $40/kW/year for both the native and wheeling customers. Equation (12.7) is used to calculate the charge rate for each wheeling customer. For the first one wheeling 50 MW from Bus DMR in the VI region to Bus KLY in the SI region, for example, the composite rate is 40 + 692/50 – 2918 × 0.5/50 + 0.217 = $24.877/kW/year. The second term is positive and reflects the load charge for the reliability degradation due to the wheeling load, and the third term is negative since it is the generation credit for the reliability improvement. For this wheeling customer, the generation credit is larger than its load charge so that its composite rate is lower than the base rate. Note that a factor of 0.5 has been applied in the third term because $2918k/year is the credit created by the 100 MW of generation at Bus DMR, whereas this wheeling customer only has the 50 MW and the other 50 MW belongs to the cogeneration customer. Similarly, the composite rates for the other two wheeling customers from Bus GMS (in the PR region) to Bus SAT (in the VI region) and from Bus REV (in the SI region) to Bus ING (in the LM region) are calculated. These are $62.137/kW/year and $58.457/kW/year, respectively. For the second and third wheeling customers, their composite rates are higher than the base rate since their generation credits are lower than the load charges. The wheeling route of the first wheeling customer is against the direction from the generation center to the load center, leading to improvement in the overall system reliability so that it should be charged less. The other two wheeling customers have the wheeling route that is along with the direction from the generation center to the load center, creating additional system risk; thus they should be charged more.

For the cogeneration customer that has the pure generator of 50 MW at Bus DMR, the credit is 2918 × 0.5 = $1459 k/year.

Equation (12.8) is used to calculate the composite charge rate for the native customers. In the calculation, EL = 7537 MW, IL = 151 MW, the increased load charge for the reliability degradation is $3605/year, and LCRGC is $0.217/kW/year. The composite rate is CR = 40 + (3605)/(7537 + 151) + 0.217 = $40.686/kW/year. The composite rate for the native customers is just slightly higher than the base rate. This is because each native customer is assumed to have only 2% of load growth

Table 12.3 Annual charges/credits and composite charge rates due to the reliability component

Customer	Generator credit (k$/yr)	Load charge (k$/yr)	Net charge (k$/yr)	Composite charge rate ($/kW/yr)
Wheeling 50 MW (DMR to KLY)	1459	692	−767	24.877
Wheeling 50 MW (GMS to SAT)	317	1413	1096	62.137
Wheeling 50 MW (REV to ING)	162	1074	912	58.457
Cogenerator (50 MW at DMR)	1459	0	−1459	
Native customs	0	3605	3605	40.686

Note: Base rate = $40/KW/yr.

and the new load charge has been distributed into the per-kilowatt rate for all the old and new loads.

The total annual charges or credits due to the reliability component and the composite charge rates for the different customers are summarized in Table 12.3.

12.6 CONCLUSIONS

This chapter discusses the method used to incorporate the reliability component into transmission service pricing. It is essential to appreciate that the rate design is not a pure technical issue but is tightly related to the power market model and other business considerations. There are different pricing principles in utility's practices. There also exist different ideas about how to incorporate system reliability into the rate design. In other words, the method presented here is certainly not unique. It is an example used to address the special application of risk evaluation in transmission service pricing. Readers can develop their own approaches once they have a clear understanding on the concept of system risk and the specific power market model.

The method discussed in this chapter is based on the following basic ideas:

- The long-run incremental capital cost can be translated into a system reliability value because the capital cost for system reinforcements to meet load growth is nothing else than the investment needed to maintain system reliability after the load growth.
- The negative or positive impacts of load or generation additions on system reliability can be quantified using a transmission risk evaluation method and converted to the charge for loads or the credit for generations using the unit incremental reliability value, which is derived from the capital investment for system reinforcements.
- Each wheeling customer includes the two elements of load addition and generation addition at different locations. The impacts of the two elements on transmission system reliability can be evaluated separately.

The main advantages of the proposed method include:

- Treating all the types of transmission service customers (native, wheeling, and generation customers) equally
- Quantifying the negative or positive impacts of the different customers on system risk and providing them with different charges or credits in terms of the impacts
- Inducing all the players in the system (the utility and customers) to share the responsibility for system reliability
- Sending an incentive price signal to encourage the right siting of new generators (such as independent power producers)
- Reflecting the long-run incremental investment requirement of the utility in the service rate
- Providing flexibility in implementation since the reliability component is just an addition to the base rate that can be determined using any conventional pricing method

CHAPTER 13

PROBABILISTIC TRANSIENT STABILITY ASSESSMENT

13.1 INTRODUCTION

Deterministic transient stability criteria have been used in power system planning and operation for years. Under such criteria, a system has to withstand the "worst case" that corresponds to extreme operating conditions and most critical contingencies. In a competitive environment, utilities also need to know the risk level associated with the criteria so that they can improve their service quality based on the consumer's expectations, that is, acceptable risk and corresponding price.

This chapter discusses probabilistic transient stability assessment and its actual applications. Similar to the adequacy assessment for steady states, the probabilistic transient stability assessment has to evaluate both the probability and consequences of fault events. However, the evaluation is more complex because of the features of transient stability simulation. The probability of fault events depends on not only the fault location and type but also on the protection scheme and disturbance sequence. The consequence evaluation requires the simulation of system transient stability and impact analysis.

There are two types of studies in this area. The first one is used to evaluate the risk of the overall system due to transient instability. In this case, the probabilities of all possible faults and their impacts for given or randomly selected prefault system states are simulated to create an overall risk index. The second one is to establish the quantitative relation between the probability of system instability and a system quantity for a given fault. The system quantity can be any operation condition that affects system stability, such as the transfer capability limit on tie lines, the generation output, or the voltage level of a key generator. The information from the first type of study is used in the general estimation of system risk, whereas the in-

Risk Assessment of Power Systems. By Wenyuan Li
ISBN 0-471-63168-X © 2005 the Institute of Electrical and Electronics Engineers, Inc.

formation from the second type of study is employed in determination of the secure operation conditions that are needed at control centers.

The probabilistic modeling and simulation methods are discussed in Section 13.2, followed by the procedure for probabilistic transient stability assessment in Section 13.3. Two application examples are provided in Section 13.4.

13.2 PROBABILISTIC MODELING AND SIMULATION METHODS [29]

13.2.1 Selection of Prefault System States

A prefault system state is defined by a network topology, a generation pattern, and a system load level. These system factors may be deterministically specified or randomly selected depending on the study purpose. For example, if the time frame in consideration is relatively long so that it can cover different system states, and the purpose is to evaluate the average risk during the given period, prefault system states should be stochastically selected using their probabilities of occurrence. If the time frame is as short as one hour (such as an hourly transfer limit study or a real-time evaluation), the network topology, generation pattern, and system load level for that hour are deterministically given.

The state sampling approach described in Section 4.3.2 can be used to randomly determine the network topology and generation availability.

The system load varies from time to time. In the "worst-case" approach, only the system peak load is used. However, the maximum loading level on a particular line may not occur at the system peak point because of noncoincidence of substation loads, network constraints, and other factors. A load curve model has to be considered in probabilistic transient stability assessment. The concept of the multiple-step load model discussed in Section 5.5.3 can be used. Essentially, this is a discrete probability distribution of the system load, with each load level having a probability value. The power-flow cases for each system load level should be prepared using the bus load allocation relationship, which is based on the load curves of substations.

13.2.2 Fault Models

The uncertainty of fault events is described using the following five models.

13.2.2.1 *Probability of Fault Occurrence.* This probability is needed if the first type of study is conducted while a line fault is deterministically specified in the second type of study. Conceptually, the probability of fault occurrence on a line can be modeled using the Poisson distribution with a constant fault rate. From the Poisson distribution formula, the probability of no fault occurring in the time period t is given by

$$P_{no} = \frac{e^{-\lambda_o t}(\lambda_o t)^0}{0!} = e^{-\lambda_o t} \tag{13.1}$$

where P_{no} is the probability of no fault occurrence, λ_o the average fault rate, and t the duration considered.

The probability of a fault occurring in t is

$$P_o = 1 - e^{-\lambda_o t} \tag{13.2}$$

Obviously, Equation (13.2) is nothing else than the fault probability following the exponential distribution. In fact, the Poisson and exponential distributions are essentially the same since both are based on the constant rate assumption.

If the condition of $\lambda_o t \ll 1$ is met, Equation (13.2) can be approximated by

$$P_o = \lambda_o t \tag{13.3}$$

The average fault rate λ_o can be approximately replaced by the frequency of fault occurrence, which could be obtained from historical records, since the duration of a fault is always extremely short.

13.2.2.2 Probability of Fault Location. The fault location on a line generally does not follow a uniform distribution. A discrete probability distribution based on historical data can be applied. A line is divided into M segments and the probability of a fault occurring in the ith segment is estimated by

$$P_i = \frac{f_i}{\displaystyle\sum_{i=1}^{M} f_i} \tag{13.4}$$

where f_i is the number of the faults occurring in the ith segment.

The number of segments is varied for different lines and depends on the data available. For instance, a line can be divided into the following three segments with respect to its master end:

1. Close-end (first 20% of the line)
2. mid-line (middle 60% of the line)
3. far-end (last 20% of the line)

13.2.2.3 Probability of Fault Type. Fault types can be classified into the following categories:

- Single-phase-to-ground
- Double-phase-to-ground
- Three-phase
- Phase-to-phase

Similarly, a discrete probability distribution for the fault type can be obtained using historical fault data. The formula has the same form as Equation (13.4) with the following definition changes: P_i is the probability of the ith fault type, f_i the number of the ith fault type, and $M = 4$.

13.2.2.4 *Probability of Unsuccessful Automatic Reclosure.* Most high-voltage overhead lines are equipped with an automatic reclosure device. A "false" fault will not create any severe consequence if the automatic reclosure is successful. The second fault clearing is needed only when the automatic reclosure fails. Conceptually, the probability of unsuccessful automatic reclosure may be obtained from historical data. However, many data-collection systems do not record the information on the success of reclosure but provide a description of the cause of faults. Generally, there is a strong correlation between the cause of faults and the probability of successful automatic reclosure. For instance, a typical statistical analysis indicates that automatic reclosure is successful for over 90% of the faults caused by lightning but for only about 50% of the faults due to other causes. The probability of unsuccessful automatic reclosure can be estimated using the conditional probability concept:

$$P_{ru} = P(L)P(U/L) + P(O)P(U/O) \tag{13.5}$$

where P_{ru} is the probability of unsuccessful automatic reclosure. $P(L)$ and $P(O)$ are the probabilities of the faults due, respectively, to lightning and other causes, which can be easily obtained from historical records. $P(U/L)$ and $P(U/O)$ are the conditional probabilities of unsuccessful reclosure given that the fault is caused by lightning and other causes, respectively, which can be estimated through statistical analysis. The equation can be extended to multiple divisions of fault causes if there are sufficient data records.

13.2.2.5 *Probability Model of Fault-Clearing Time.* The process of clearing a fault is composed of three components: fault detection, relay operation, and breaker operation. The fault detection can be assumed to be complete instantly. The times required for the relay and breaker operations are both random. There are two modeling approaches for the fault-clearing time:

1. The relay and breaker operation times are modeled using two separate probability distributions and the probability distribution of the total fault clearing time is obtained by a convolution.
2. The total fault-clearing time is assumed to directly follow a probability distribution.

Generally, the normal distribution assumption is used for the relay or breaker operation time, or the total fault-clearing time. The estimate of the mean and its standard deviation for the normal distribution can be determined from historical data. A system loses transient stability only when the fault clearing time is longer than the critical clearing time.

13.2.3 Monte Carlo Simulation of Fault Events

The impact of a fault on system transient stability depends on the following three factors:

1. The prefault system state, which is defined by the network topology, generation availability, and system load level
2. The features of the fault itself, which include the fault line, fault location, and fault type
3. The protection operation, which includes the successful or unsuccessful reclosure and fault-clearing time.

As mentioned earlier, a prefault system state may be deterministically specified or randomly selected depending on the time frame considered. Selection of fault lines depends on the study purpose. In evaluating the risk of the overall system due to transient instability (the first type of study), all line faults should be randomly simulated. In establishing the quantitative relation between the probability of system instability and an operation condition (the second type of study), a fault line is often deterministically specified. All the other factors are treated as random variables.

With the fault models in Section 13.2.2, Monte Carlo simulations of fault events can be methodologically classified into the following three categories:

1. Sampling with two possibilities. This category includes determination of a fault occurrence or a reclosure failure. Consider the fault occurrence as an example. A uniformly distributed random number R between $[0, 1]$ is created. If $R < P_o$, the fault occurs; otherwise the fault does not occur. This is shown in Figure 13.1.
2. Sampling with multiple possibilities. This category includes determination of a fault location or a fault type. Consider the fault type as an example. The probability values of the four fault types are successively placed between $[0, 1]$, as shown in Figure 13.2. A uniformly distributed random number R between $[0, 1]$ is created. The location of R denotes which fault type is randomly selected in the sampling.
3. Sampling with a normally distributed random variable. The fault-clearing time sampling belongs to this category, which includes the following two steps:
   - Create a standard normally distributed random number X using the method described in Appendix B.4.2.
   - The random fault-clearing time is calculated by

$$\tau_c = X\sigma + \mu \tag{13.6}$$

where μ and σ are, respectively, the mean and standard deviation of the fault-clearing time.

Figure 13.1 Sampling fault occurrence.

13.2.4 Transient Stability Simulation

For a fault event, the transient stability simulation is performed to determine whether or not it causes system instability. The time domain simulation method is used. Generally, the number of transient stability simulations is very large since system states and fault events are randomly selected and a great number of samples are necessary to produce a probabilistic risk index or its probability distribution. Many techniques have been developed to speed up the simulation process, such as the variable-step algorithm for differential equations, the early-termination criterion, the fast contingency-screening approach, and the second-kick method.

It is important to appreciate that a disturbance sequence must be specified in the transient stability simulation. The disturbance sequence is varied for different fault events and is usually defined in the operation criteria of a control center, which are based on considerable off-line investigation of system transient stability. In addition to the relay and breaker operations and automatic reclosure, the disturbance sequence also includes various remedial actions, such as generation rejection, switching of Var equipment, intentional tripping of lines, and actions of control systems.

13.3 PROCEDURE

The procedure for probabilistic transient stability assessment is somewhat different for the two types of studies.

13.3.1 Procedure for the First Type of Study

The procedure for evaluating the risk of the overall system due to transient instability is shown in Figure 13.3. In the "select" blocks, the network topology and

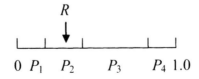

Figure 13.2 Sampling fault type.

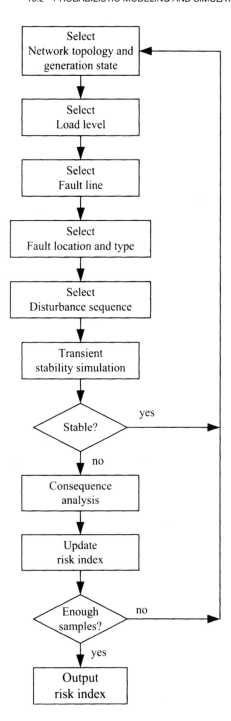

Figure 13.3 Procedure for evaluating overall system risk due to transient instability.

generation states may be randomly or deterministically selected depending on the time length of the study; all the other selections are performed using the Monte Carlo simulation. The risk index is a combination of consequences of fault events and their probabilities of occurrence. The block labeled "consequence analysis" in the figure is the evaluation of the damage cost due to transient instability. It is system-state and fault-event specific and is associated with the following aspects:

- Load shedding due to generation rejections
- Equipment damages
- Transfer limit reductions that may lead to losses of revenue
- Cascading failures that may cause a catastrophic outcome
- Penalties for criteria violation

13.3.2 Procedure for the Second Type of Study

The procedure for establishing the quantitative relation between the probability of system instability and a system condition for a given fault is shown in Figure 13.4. This type of study has the following features:

- The fault is specified.
- The system parameters are adjusted to make the system stable so that a transient stability limit can be established.
- The output is the relation between the probability of instability and a key operation condition.

In the figure, the term "barely stable" means the case in which increasing the value of a stability parameter by a small increment will result in an unstable state. For instance, if a stability parameter has the threshold accuracy of 1 MW, then a case is barely stable if it is a stable case but will become unstable by increasing the stability parameter by 1 MW.

13.4 EXAMPLES [29]

Two examples in a utility system are given in this section to demonstrate the application.

13.4.1 System Description and Data

The bulk system in the study is composed of two main generation subsystems: the Peace River and Columbia River systems. The Peace River system has a generating capacity of 3400 MW and is located about 1000 km away from the main load cen-

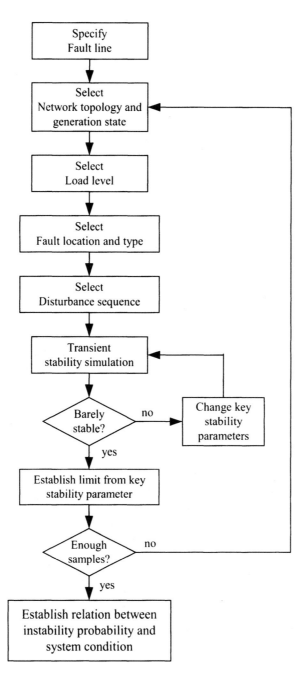

Figure 13.4 Procedure for establishing relation between instability probability and system condition.

ter, and the Columbia River system has a generating capacity of 4730 MW and is located 200 to 500 km from the load center. The large transfer flows over long distances characterize the utility system as one limited by transient stability, with generation shedding used as an effective remedial action for maintaining stability.

The purpose of the study was to establish the relation between the probability of system instability and system operation conditions in the 1997 summer season. To demonstrate the impact on the calculated stability limits, probabilistic transient stability assessments were conducted as part of a transfer limit study on the Columbia River system and a generation rejection study on the Peace River system. The results are compared with the limits obtained using the traditional deterministic methods.

The main data are given as follows.

13.4.1.1 System Load Data. Based on the actual hourly load records in the 1995/1996 fiscal year and the load forecast for the 1997 summer season, a six-step model was created and is shown in Table 13.1. The load steps formed the basis for developing the different prefault power flow cases used in the studies.

13.4.1.2 Fault Model Data. The fault statistics on the 500 kV lines from 1977 to 1996 were retrieved from the utility database. During this time period, 2419 faults were recorded.

A 500 kV transmission line was divided into three parts with respect to its master end: the close-end (first 20% of the line), mid-line (middle 60%), and far-end (last 20%). The historical records of the faults were used to calculate the fault location probabilities as shown in Table 13.2.

Of the 2419 fault events recorded, 43 did not have the fault type noted. Although these cases could have been proportionally allocated among the other fault types, it was decided to pessimistically consider the unknown events as three-phase faults. Table 13.3 shows the probabilities of the fault types.

In reviewing the cause of the historical fault events, it was noted that there was a very high correlation between the cause of the faults and the probability of successful automatic reclosure. The automatic reclosure was successful in about 90% of the faults caused by lightning but in about 50% of the faults due to all other causes. According to the historical records, 82.51% of the faults are due to lightning and

Table 13.1 System load model

Load level (%)	Load (MW)	Probability
100	7420.2	0.0263
95	7049.2	0.2192
90	6678.2	0.1504
85	6307.2	0.1458
80	5936.2	0.2034
70	5194.1	0.2549

Table 13.2   Fault location model

Fault location	Probability
Close-end	0.1307
Middle-line	0.7021
Far-end	0.1672

Table 13.3   Fault type model

Fault type	Probability
Single-phase-to-ground	0.8851
Double-phase-to-ground	0.0438
Phase-to-phase	0.0401
Three-phase	0.0132
Unknown	0.0178

17.49% due to the other causes. The probability of unsuccessful automatic reclosure is thus estimated as:

$$P_{ru} = 0.8251 \times (1 - 0.9) + 0.1749 \times (1 - 0.5) = 0.17$$

The probability of reclosure operation is listed in Table 13.4.

In the traditional deterministic studies, a fixed mean value of 4 cycles for 500 kV line fault clearing was used. In the probabilistic transient stability assessment, the probabilities of breaker and relay operation times were simulated. It was found from the investigations that the breaker operating time was fixed at 1.5 or 2.0 cycles, depending on which side of the breaker was opened first. Since no other data on the breaker operation was available, it was assumed that the probability of breaker operation was evenly split between the two times. It was decided that the relay operating time could be modeled using the normal distribution with a mean of 2 cycles and a 10% standard deviation. The fault detection was assumed to occur instantaneously.

13.4.2   Transfer Limit Calculation in the Columbia River System

The first study was performed for a fault on a 500 kV line close to the Columbia River generation system. In the study, the transfer flow was adjusted to make the

Table 13.4   Reclosure model

Reclosure	Probability
Successful	0.83
Unsuccessful	0.17

system stable. The sequence of the events for this contingency is given below. The R is a random value that is determined using the Monte Carlo simulation on the clearing time.

Time (cycle)	Event
0	fault (various fault locations and types)
2	series capacitor control
R	clearing fault
9	tripping generator to maintain stability
60	reclosure of fault at the master end
$60 + R$	clearing fault if unsuccessful measure

For the given study conditions, one thousand sample cases were generated, for all of which the stability limits were calculated. Comparative studies for the identical contingency using the traditional deterministic criteria of the utility revealed that a reduction of 640 MW in the transfer capability was required to maintain system transient stability. By contrast, the numerical results from the probabilistic studies are given in Table 13.5 and Figure 13.5. The results, which provide the distribution of the transfer limit reduction required to maintain stability for all the 1000 cases, indicate that in 847 of the 1000 cases, no reduction in the transfer capacity was required to maintain system stability. The maximum amount of the reduction required was 780 MW, which has been rounded to 800 MW in the table. It is inter-

Table 13.5 Probability of transfer limit reduction

Transfer limit reduction (MW)	Number of cases	Cumulative probability of stability (%)
0	847	84.70
50	1	84.80
100	0	84.80
150	0	84.80
200	32	88.00
250	1	88.10
300	14	89.50
350	38	93.30
400	19	95.20
450	0	95.20
500	7	95.90
550	31	99.00
600	9	99.90
650	0	99.90
700	0	99.90
750	0	99.90
800	1	100.00
more	0	100.00

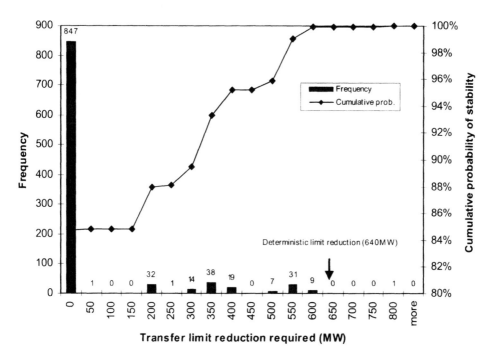

Figure 13.5   Distribution of transfer limit reduction required to maintain system stability.

esting to note that this amount was required at the system load level of 95% of the peak (not 100% as might be assumed) and was higher than the 640 MW required when the traditional deterministic criteria (i.e., 100% load level, four-cycle clearing, unsuccessful reclosure) was applied. This result is due to the fact that the amount of power being transmitted over the faulted line at the 95% system load level is higher than that when the system load level is 100%.

Figure 13.6 shows the probability of instability associated with each transfer limit reduction. The results indicate that a transfer limit reduction of 800 MW and more corresponds to 0% risk of instability. The results also indicate that the traditional deterministic criteria are slightly conservative and still have to accept the 0.1% probability of instability. For the specific contingency studied, the regular practice would be to reduce exports by about 640 MW at the high system load level. It can be seen from the results of probability distribution that if the company were willing to accept just a 0.1% chance of instability, then the transfer limit should be reduced by about 600 MW. If a 1.0% chance of instability is acceptable, the transfer limit would be reduced only by about 550 MW.

13.4.3   Generation Rejection Requirement in the Peace River System

The second study was performed for a fault on a 500 kV line close to the Peace River generation system. The switching sequence for the contingency is given be-

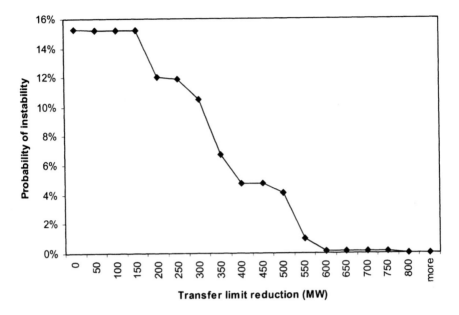

Figure 13.6   Transfer limit reduction versus probability of instability.

low. The R is a random value of the clearing time obtained using the Monte Carlo simulation.

Time (cycle)	Event
0	fault (various fault locations and types)
2	bypassing the series capacitor if the rating is exceeded
R	clearing fault
9	switching 200 to 400 MW of braking resistors on tripping generator to maintain stability
35	reclosure of fault at the master end
35 + R	clearing fault if unsuccessful reclosure
44	switching braking resistors off

For the above study conditions, one thousand sample cases were generated, for all of which the stability limits were calculated. Comparative studies for the identical contingency using the traditional deterministic criteria revealed that a total of 1490 MW of generation shedding was required to maintain system transient stability. By contrast, the numerical results from the probabilistic studies are given in Table 13.6 and Figure 13.7. The results, which provide the distribution of the required generation shedding to maintain stability for all the 1000 cases, show that in 821 of the 1000 cases, no generation shedding was required to maintain system stability. The maximum amount of shedding required was 1240 MW, which has been

Table 13.6 Probability of generation shedding

Generation shedding (MW)	Number of cases	Cumulative probability of stability (%)
0	821	82.10
100	14	83.50
200	74	90.90
300	39	94.80
400	2	95.00
500	36	98.60
600	2	98.80
700	3	99.10
800	3	99.40
900	3	99.70
1000	1	99.80
1100	0	99.80
1200	0	99.80
1300	2	100.00
1400	0	100.00
more	0	100.00

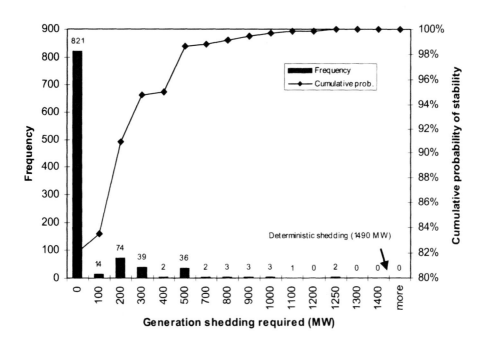

Figure 13.7 Distribution of generation shedding required to maintain system stability.

rounded to 1300 MW in the table as a conservative rounding. However, it should be noted that the usual deterministic case of 100% load level and three-phase permanent (unsuccessful reclosure) fault did not appear in the 1000 cases selected using the Monte Carlo technique. If that particular case had been selected, the maximum amount of shedding required would also have been 1490 MW.

Figure 13.8 shows the probability of instability for each level of generation shedding. The results indicate that the traditional deterministic criteria are very conservative. For the specific contingency studied, the regular practice would be to arm the generation shedding scheme to shed 1490 MW of generation. This rejected generation would lead to the loss of export sales of the same amount. The case of 1490 MW of generation shedding is not selected in the Monte Carlo sampling so that its probability has been shown as 0% in Table 13.6 and Figure 13.8. The actual probability for this case is very small (between 0% and 0.1%). The remaining results indicate that if the company were willing to accept just a 0.2% chance of instability, then the generation shedding scheme should be armed to shed about 1000 MW in the event where this particular contingency occurs. Accepting a probability of instability of 1.4% would reduce this figure by about 500 MW.

13.4.4 Summary

The two examples have shown that the probabilistic transient stability assessment provides a wider understanding of the physical system limits. The traditional deter-

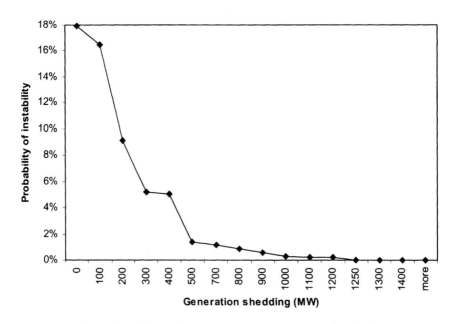

Figure 13.8 Generation shedding versus probability of instability.

ministic criterion employed by the utility has been shown to be conservative, with a probability of instability between 0% and 0.1% in the Peace River case and 0.1% in the Columbia River case.

The traditional deterministic criteria are applied partly because of the belief that the system should survive the "worst-case" contingency. However, the deterministic criteria do not always correspond to the "worst case." As seen in the Columbia River study, the worst case for the contingency considered was actually at the 95% system load level, not at the 100% system load level that is normally used in the deterministic analysis. Line flows depend not only on the total system load level but also on the noncoincidence of substation loads, generation patterns, reactive sources, and other factors. In other words, the "worst case" may actually be missed in applying the deterministic criteria and a small probability of instability may be still unintentionally accepted.

In the study on the Peace River system, the "worst case," in which the 1490 MW generation shedding should be required to maintain system transient stability, was missed from the 1000 samples. This suggests that a larger number of samples are needed to achieve more accurate results. Generally, the coefficient of variance should be used as the stopping rule in the Monte Carlo simulation.

13.5 CONCLUSIONS

Probabilistic transient stability assessment requires two simulations: randomness of fault events and system stability. The randomness of fault events is associated with selection of the prefault system states defined by the network topology, generation availability, system load level, determination of faults (including the fault line, location, and type), and probabilistic modeling of the protection schemes (including the automatic reclosure, relay, and breaker operations). Apparently, this is a much more complex task compared to the steady-state system risk evaluation in which only system states are randomly selected. The Monte Carlo sampling technique is superior to the state enumeration method in thhis type of randomness simulation. The system stability simulation should include not only the power system network but also protection operations, control systems, and remedial actions. Different disturbance sequences for individual faults need to be specified and simulated. A time domain tool for stability simulation should be used. This is a time-consuming process but it provides accuracy and flexibility in system modeling. Various approaches to speed the transient stability simulation should be considered.

Probabilistic transient stability assessment can be classified into two categories in terms of the study purpose. The first one is the risk evaluation of the overall system due to transient instability. In this case, the probabilities of the faults on all critical lines and their impacts are simulated to create an overall risk index. The second one is the calculation of the probability distribution of system instability versus a system operation condition for a given fault. In the stability simulation for the second category, the parameter representing the system operation condition is adjusted to make the system stable, creating a critical stability limit.

The prefault system states can be stochastically or deterministically selected in a probabilistic transient stability assessment depending on the time frame for the study. If a relatively long time period is considered, a variety of possible prefault system states should be randomly determined to obtain the average indices. If the time frame in an assessment is one hour or less, such as in a real-time application, a prefault system state is deterministically specified.

The benefit/cost analysis can be conducted to establish some new operation criteria based on the risk assessment of transient instability. The basic concept is similar to that discussed in the previous chapters and has not been detailed in this chapter. It is important to recognize that the purpose of using the probabilistic transient stability assessment is not to replace the traditional deterministic criteria, which have been utilized for years in the power industry. However, the probabilistic method provides a sophisticated complement to enhance the transient stability analysis in utilities.

APPENDIX A

BASIC PROBABILITY CONCEPTS

A.1 PROBABILITY CALCULATION RULES

A.1.1 Intersection

Given two events A and B, the probability of simultaneous occurrence of A and B is calculated by

$$P(A \cap B) = P(A)(B/A) \tag{A.1}$$

where $P(B/A)$ is the conditional probability of B occurring given that A has occurred. If A and B are independent, Equation (A.1) becomes

$$P(A \cap B) = P(A)P(B) \tag{A.2}$$

This can be generalized to the case of N events. If N events are independent of each other, the following probability equation holds:

$$P(A_1 \cap A_2 \cap \ldots \cap A_N) = P(A_1)P(A_2) \ldots P(A_N) \tag{A.3}$$

A.1.2 Union

Given two events A and B, the probability of occurrence of either A or B or both is calculated by

$$P(A \cup B) = P(A) + P(B) - P(A \cap B) \tag{A.4}$$

Risk Assessment of Power Systems. By Wenyuan Li
ISBN 0-471-63168-X © 2005 the Institute of Electrical and Electronics Engineers, Inc.

If A and B are mutually exclusive, $P(A \cap B) = 0$. This can be generalized to the case of N events. If N events are mutually exclusive, the following probability equation holds:

$$P(A_1 \cup A_2 \cup \ldots \cup A_N) = P(A_1) + P(A_2) + \ldots + P(A_N) \tag{A.5}$$

A.1.3   Full Conditional Probability

If the events $\{B_1, B_2, \ldots, B_N\}$ represent a full and mutually exclusive set, that is, $P(B_1) + P(B_2) + \cdots + P(B_N) = 1.0$ and $P(B_i \cap B_j) = 0.0$ $(i \neq j; i, j = 1, 2, \ldots, N)$, then for any event A,

$$P(A) = \sum_{i=1}^{N} P(B_i)P(A/B_i) \tag{A.6}$$

This equation is often used in the case of two conditional events. If B_1 and B_2 are mutually exclusive and $P(B_1) + P(B_2) = 1.0$, then

$$P(A) = P(B_1)P(A/B_1) + P(B_2)P(A/B_2) \tag{A.7}$$

A.2   RANDOM VARIABLE AND ITS DISTRIBUTION

A random event can be represented using a random variable. Given a continuous random variable X, the probability of X being not larger than a real number x is a function of x. This function is called the cumulative distribution function $F(x)$ of the random variable X, which can be expressed in the following form:

$$F(x) = \int_{-\infty}^{x} f(x)dx \tag{A.8}$$

where $f(x)$ is the probability density function. Obviously,

$$f(x) = \frac{dF(x)}{dx} \tag{A.9}$$

The probability of X lying between a and b is calculated by

$$P(a \leq X \leq b) = \int_{a}^{b} f(x)dx \tag{A.10}$$

For a discrete random variable, its probability density function can be expressed as

$$p_k = P(X = x_k) \qquad (k = 1, 2, \ldots) \tag{A.11}$$

and its cumulative probability distribution function as

$$F(x_k) = P(X \le x_k) \qquad (k = 1, 2, \ldots) \tag{A.12}$$

The relationship between the density and cumulative distribution functions of a discrete random variable is given by

$$F(x_k) = \sum_{i \le k} p_i \tag{A.13}$$

and

$$p_k = F(x_k) - F(x_{k-1}) \tag{A.14}$$

A.3   IMPORTANT DISTRIBUTIONS IN RISK EVALUATION

The following are the four important distributions in risk evaluation:

A.3.1   Exponential Distribution

This is a single-parameter distribution. The density function is

$$f(x) = \lambda \exp(-\lambda x) \qquad (x \ge 0) \tag{A.15}$$

The cumulative distribution function is

$$F(x) = 1 - \exp(-\lambda x) \tag{A.16}$$

$$= \lambda x - \frac{(\lambda x)^2}{2!} + \frac{(\lambda x)^3}{3!} - \cdots$$

When $\lambda x \ll 1$, Equation (A.16) is approximated by

$$F(x) = \lambda x \tag{A.17}$$

The mean and variance of the exponential distribution are $1/\lambda$ and $1/\lambda^2$, respectively.

A.3.2   Normal Distribution

The density function is

$$f(x) = \frac{1}{\sigma\sqrt{2\pi}} \exp\left[-\frac{(x - \mu)^2}{2\sigma^2}\right] \qquad (-\infty \le x \le \infty) \tag{A.18}$$

where μ and σ^2 are the mean and variance of the normal distribution.

By using the following substitution,

$$z = \frac{x - \mu}{\sigma} \qquad \text{(A.19)}$$

Equation (A.18) becomes

$$f(z) = \frac{1}{\sqrt{2\pi}} \exp\left[-\frac{z^2}{2}\right] \qquad (-\infty \leq z \leq \infty) \qquad \text{(A.20)}$$

Equation (A.20) is the density function of the standard normal distribution.

There is no explicitly analytical expression for the cumulative distribution function of the normal distribution. The area $Q(z)$ under the standard normal density function curve shown in Figure A.1 can be found from the following polynomial approximation for $z \geq 0$ [2]:

$$Q(z) = f(z)[b_1 t + b_2 t^2 + b_3 t^3 + b_4 t^4 + b_5 t^5] \qquad \text{(A.21)}$$

where $t = \dfrac{1}{1 + rz}$

$$r = 0.2316419$$

$$b_1 = 0.31938153$$

$$b_2 = -0.356563782$$

$$b_3 = 1.781477937$$

$$b_4 = -1.821255978$$

$$b_5 = 1.330274429$$

The maximum error of Equation (A.21) is smaller than 7.5×10^{-8}.

A.3.3 Log-Normal Distribution

The density function is

$$f(x) = \frac{1}{x\sigma\sqrt{2\pi}} \exp\left[-\frac{(\ln x - \mu)^2}{2\sigma^2}\right] \qquad (x > 0) \qquad \text{(A.22)}$$

It is important to recognize that μ and σ^2 in Equation (A.22) are not the mean and variance of the log-normal distribution. The mean and variance of the log-normal distribution are given, respectively, by

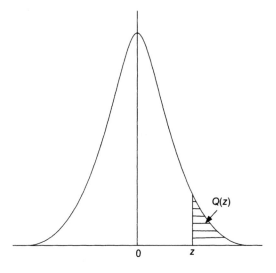

Figure A.1   Area under the standard normal density function.

$$E(x) = \exp\left(\mu + \frac{\sigma^2}{2}\right) \tag{A.23}$$

$$V(x) = \exp(2\mu + \sigma^2)[\exp(\sigma^2) - 1] \tag{A.24}$$

If the mean and variance of the log-normal distribution are prespecified, the parameters μ and σ^2 in Equation (A.22) are calculated by

$$\mu = \ln\left[\frac{E^2}{(V + E^2)^{1/2}}\right] \tag{A.25}$$

$$\sigma^2 = \ln\left[\frac{V + E^2}{E^2}\right] \tag{A.26}$$

A.3.4   Weibull Distribution

The density function is

$$f(x) = \frac{\beta x^{\beta-1}}{\alpha^\beta} \exp\left[-\left(\frac{x}{\alpha}\right)^\beta\right] \qquad (\infty > x \geq 0,\ \beta > 0,\ \alpha > 0) \tag{A.27}$$

The cumulative distribution function is

$$F(x) = 1 - \exp\left[-\left(\frac{x}{\alpha}\right)^\beta\right] \qquad (\infty > x \geq 0,\ \beta > 0,\ \alpha > 0) \tag{A.28}$$

The mean and standard deviation of the Weibull distribution can be calculated from the scale (α) and shape (β) parameters as follows:

$$\mu = \alpha\Gamma\left(1 + \frac{1}{\beta}\right) \tag{A.29}$$

$$\sigma^2 = \alpha^2\left[\Gamma\left(1 + \frac{2}{\beta}\right) - \Gamma^2\left(1 + \frac{1}{\beta}\right)\right] \tag{A.30}$$

where $\Gamma(\bullet)$ is the gamma function which is defined by

$$\Gamma(x) = \int_0^\infty t^{x-1}e^{-t}dt \tag{A.31}$$

There is no analytical expression for α or β using the mean (μ) and standard deviation (σ). The following method can be used to calculate α and β from μ and σ. By eliminating α from Equations (A.29) and (A.30), we have

$$\frac{\Gamma\left(1 + \dfrac{2}{\beta}\right)}{\Gamma^2\left(1 + \dfrac{1}{\beta}\right)} = 1 + \frac{\sigma^2}{\mu^2} \tag{A.32}$$

Using an approximate expression of the gamma function [79], Equation (A.32) is approximated by

$$\frac{\left(1 + \dfrac{2}{\beta}\right)^{(0.5+2/\beta)} e^{-(1+2/\beta)}\left[1 + 1/12\left(1 + \dfrac{2}{\beta}\right)\right]}{\left[\left(1 + \dfrac{1}{\beta}\right)^{(1+2/\beta)} e^{-(2+2/\beta)}\left[1 + 1/12\left(1 + \dfrac{1}{\beta}\right)\right]\right]^2 \sqrt{2\pi}} = 1 + \frac{\sigma^2}{\mu^2} \tag{A.33}$$

Equation (A.33) can be solved to obtain β using a bifurcation algorithm. Then α is found from (A.30) using β and σ.

A.4   NUMERICAL CHARACTERISTICS

Random variables can be simply described using one or more parameters that are called numerical characteristics. The most useful numerical characteristics in risk evaluation are the mathematical expectation (mean), variance or standard deviation, covariance, and correlation coefficients. As a matter of fact, a risk index is a mean value, whereas a standard deviation of an estimate is often used as an indicator of the accuracy in Monte Carlo simulation.

A.4.1 Mathematical Expectation

If a random variable X has the probability density function $f(x)$ and the random variable Y is a function of X, that is, $y = y(x)$, then the mathematical expectation or mean value of Y is defined as

$$E(Y) = \int_{-\infty}^{\infty} y(x)f(x)dx \tag{A.34}$$

As a special case of the general definition, the mean of the random variable X is

$$E(x) = \int_{-\infty}^{\infty} xf(x)dx \tag{A.35}$$

For a discrete random variable, Equations (A.34) and (A.35) become Equations (A.36) and (A.37):

$$E(Y) = \sum_{i=1}^{n} y(x_i)p_i \tag{A.36}$$

$$E(X) = \sum_{i=1}^{n} x_i p_i \tag{A.37}$$

A.4.2 Variance and Standard Deviation

For a random variable X with the probability density function $f(x)$, its variance is defined as:

$$V(X) = E\{[X - E(X)]^2\} = \int_{-\infty}^{\infty} [x - E(X)]^2 f(x)dx \tag{A.38}$$

If X is a discrete random variable, Equation (A.38) becomes

$$V(X) = \sum_{i=1}^{n} [x_i - E(X)]^2 p_i \tag{A.39}$$

The variance is an indicator for the dispersion degree of possible values of X from its mean. The square root of the variance is known as the standard deviation and is often expressed by the notation $\sigma(X)$.

A.4.3 Covariance and Correlation Coefficients

Given an N-dimension random vector $(X_1, X_2, \ldots, X_N)$, the covariance between any two elements X_i and X_j is defined as

$$c_{ij} = E\{[X_i - E(X_i)][X_j - E(X_j)]\}$$
$$= E(X_iX_j) - E(X_i)E(X_j)$$

(A.40)

The covariance is often expressed using the notation $\text{cov}(X_i, X_j)$. The covariance between an element and itself is its variance:

$$\text{cov}(X_i, X_i) = V(X_i)$$

(A.41)

The correlation coefficient of X_i and X_j is defined as

$$\rho_{ij} = \frac{\text{cov}(X_i, X_j)}{\sqrt{V(X_i)}\sqrt{V(X_j)}}$$

(A.42)

The absolute value of ρ_{ij} is smaller or equal to 1.0. If $\rho_{ij} = 0$, X_i and X_j are not correlated; if $\rho_{ij} > 0$, X_i and X_j are positively correlated; and if $\rho_{ij} < 0$, X_i and X_j are negatively correlated.

APPENDIX B

ELEMENTS OF MONTE CARLO SIMULATION

B.1 GENERAL CONCEPT

The basic idea of Monte Carlo simulation is to create a series of experimental samples using a random number sequence. According to the Central Limit Theorem or the Law of Large Numbers, the sample mean can be used as an unbiased estimate of the mathematical expectation when the number of samples is large enough. The variance of the sample mean is an indicator of estimation accuracy.

Take the application of Monte Carlo simulation to risk assessment as an example. Let U denote the unavailability or failure probability of a system and x_i be the zero–one indicator variable that can be obtained using a Monte Carlo simulation method:

$x_i = 1$ if the sampled system state is a failed one
$x_i = 0$ if the sampled system state is a successful one

The estimate of the system unavailability is given by the following sample mean:

$$U = \frac{1}{N} \sum_{i=1}^{N} x_i \qquad (\text{B.1})$$

where N is the number of system state samples.

The sample variance is defined as

$$V(x) = \frac{1}{N-1} \sum_{i=1}^{N} (x_i - U)^2 \qquad (\text{B.2})$$

Risk Assessment of Power Systems. By Wenyuan Li
ISBN 0-471-63168-X © 2005 the Institute of Electrical and Electronics Engineers, Inc.

The estimated unavailability given in Equation (B.1) is also a random variable that depends on the number of samples and the sampling process. The uncertainty around the estimate can be measured by the variance of the sample mean, which is defined as

$$V(U) = \frac{V(x)}{N} = \frac{1}{N(N-1)} \sum_{i=1}^{N} (x_i - U)^2 \tag{B.3}$$

It can be seen that the sample variance $V(x)$ given by Equation (B.2) and the variance of the sample mean $V(U)$ given by Equation (B.3) are two different concepts and should not be confused with each other.

The standard deviation of the sample mean is

$$\sigma = \sqrt{V(U)} = \frac{\sqrt{V(x)}}{\sqrt{N}} \tag{B.4}$$

Equation (B.4) indicates that two measures can be used to reduce the standard deviation of the estimate in Monte Carlo simulation: increasing the number of samples or decreasing the sample variance. Many variance reduction techniques have been developed to improve the effectiveness of Monte Carlo simulation. It is important to appreciate that the variance cannot be reduced to zero in any case and, therefore, it is always necessary to consider a reasonable and sufficiently large number of samples.

Monte Carlo simulation creates a fluctuating convergence process and there is no guarantee that a few more samples will definitely lead to a smaller error. It is true, however, that the error bound or confidence range decreases as the number of samples increases. The accuracy level of Monte Carlo simulation can be measured using the coefficient of variance, which is defined as the standard deviation of the estimate divided by the estimate:

$$\eta = \sqrt{V(U)}/U \tag{B.5}$$

The coefficient of variance is often used as a convergence criterion.

B.2 RANDOM NUMBER GENERATORS

Generating a random number is a key step in Monte Carlo simulation. Theoretically, a random number generated by a mathematical method is not really random and is called a pseudorandom number. In principle, a pseudorandom number sequence should be statistically tested to assure its randomness.

B.2.1 Multiplicative Congruent Generator

The multiplicative congruent generator is given by the following recursive relationship:

$$x_{i+1} = ax_i (\text{mod } m) \tag{B.6}$$

where a is the multiplier and m is the modulus; a and m have to be nonnegative integers. The module notation (mod m) means that

$$x_{i+1} = ax_i - mk_i \tag{B.7}$$

where $k_i = [ax_i/m]$ denotes the largest positive integer in ax_i/m.

Given an initial value x_0 that is called a seed, Equation (B.6) generates a random number sequence which lies between $[0, m]$. A random number sequence uniformed distributed in the interval $[0, 1]$ can be obtained by

$$R_i = \frac{x_i}{m} \tag{B.8}$$

Obviously, the random number sequence generated using Equation (B.6) will repeat itself in at most m step and is periodic. If the repeat period equals m, it is called a full period. Different choices of the parameters a and m produce large impacts on the statistical features of random numbers. Based on many statistical tests, the following parameters provide satisfactory statistical features in generated random numbers:

$m = 2^{31} - 1$

$a = 16807$ or 630360016

The initial value x_0 is any odd number.

B.2.2 Mixed Congruent Generator

The mixed congruent generator is given by the following recursive relationship:

$$x_{i+1} = (ax_i + c)(\text{mod } m) \tag{B.9}$$

Compared to the multiplicative congruent generator, the new parameter c is added in the mixed congruent generator. The quantity c is called the increment and it also has to be a nonnegative integer. Statistical tests indicate that the following two sets of parameters provide satisfactory statistical features in generated random numbers:

$m = 2^{31}$

$a = 314159269$

$c = 453806245$

$m = 2^{35}$

$a = 5^{15}$

$c = 1$

B.3   INVERSE TRANSFORM METHOD OF GENERATING RANDOM VARIATES

A random variate refers to a random number sequence following a given distribution. The two random number generators presented in Section B.2 generate the random number sequence following a uniform distribution between [0, 1]. The methods for generating the random variates following other distributions can be classified into three categories: (1) inverse transform method, (2) composition method, and (3) acceptance–rejection method. This section only covers the inverse transform method, which is most commonly used.

The inverse transform method is based on the following proposition:

If a random variate R follows a uniform distribution in the interval between [0, 1], the random variate $X = F^{-1}(R)$ has a continuous cumulative probability distribution function $F(x)$.

The proposition can be generalized to the case of a discrete distribution, and in this case the inverse function is defined as

$$X = F^{-1}(R) = \min\{x : F(x) \geq R\} \qquad (0 \leq R \leq 1) \qquad (B.10)$$

The procedure of generating random variates using the inverse transform method is as follows:

- Generate a uniformly distributed random number sequence R between [0, 1]
- Calculate the random variate which has the cumulative probability distribution function $F(x)$ by $X = F^{-1}(R)$

For some random variates, the cumulative probability distribution function does not have the explicit expression of its inverse function. In this case, an approximate analytical expression is needed.

B.4   IMPORTANT RANDOM VARIATES IN RISK EVALUATION

B.4.1   Exponential Distribution Random Variate

The cumulative probability distribution function of the exponential distribution is

$$F(x) = 1 - e^{-\lambda x} \qquad (B.11)$$

A uniform distribution random number R is generated so that

$$R = F(x) = 1 - e^{-\lambda x} \qquad (B.12)$$

Using the inverse transform method, we have

$$X = F^{-1}(R) = -\frac{1}{\lambda} \ln(1 - R) \tag{B.13}$$

Since $(1 - R)$ distributes uniformly in the same way as R in the interval $[0, 1]$, Equation (B.13) equivalently becomes

$$X = -\frac{1}{\lambda} \ln(R) \tag{B.14}$$

where R is a uniform distribution random number sequence and X follows the exponential distribution.

B.4.2 Normal Distribution Random Variate

There exists no analytical expression for the inverse function of the normal cumulative distribution function. The following approximate expression can be used. Given an area $Q(z)$ under the normal density distribution curve as shown in Fig. A.1, the corresponding z can be calculated by [2]

$$z = s - \frac{\sum\limits_{i=0}^{2} c_i s^i}{1 + \sum\limits_{i=1}^{3} d_i s^i} \tag{B.15}$$

where

$$s = \sqrt{-2 \ln Q} \tag{B.16}$$

and

$c_0 = 2.515517$
$c_1 = 0.802853$
$c_2 = 0.010328$
$d_1 = 1.432788$
$d_2 = 0.189269$
$d_3 = 0.001308$

The maximum error of Equation (B.15) is smaller than 0.45×10^{-4}.

The algorithm for generating the normal distribution random variate is stated as follows:

Step 1: Generate a uniform distribution random number sequence R between $[0, 1]$.

Step 2: Calculate the normal distribution random variate X by

$$X = \begin{cases} z & \text{if } 0.5 < R \le 1.0 \\ 0 & \text{if } R = 0.5 \\ -z & \text{if } 0 \le R < 0.5 \end{cases} \tag{B.17}$$

where z is obtained from Equation (B.15) and Q in Equation (B.16) is given by

$$Q = \begin{cases} 1 - R & \text{if } 0.5 < R \le 1.0 \\ R & \text{if } 0 \le R \le 0.5 \end{cases} \tag{B.18}$$

B.4.3   Log-Normal Distribution Random Variate

According to the basic probability theory, if the random variable Y is a function of the random variable X, that is, $y = y(x)$, then the probability density functions of Y and X have the following relationship:

$$f(y) = f(x) \left| \frac{dx}{dy} \right| \tag{B.19}$$

It can be proved from Equation (B.19) that if X follows the normal distribution, then $Y = e^X$ follows the log-normal distribution.

The algorithm for generating the log-normal distribution random variate is as follows:

Step 1: Generate a random variate Z following the standard normal distribution.

Step 2: Let $X = \mu + \sigma Z$, where μ and σ are the parameters in the density function of the log-normal distribution. Note that as mentioned in Section A.3.3, they are not the mean and standard deviation of the log-normal distribution.

Step 3: Let $Y = e^X$, where Y is a log-normally distributed random variate.

B.4.4   Weibull Distribution Random Variate

By using the inverse transform method, let a uniform distribution random number R equal the Weibull cumulative probability distribution function, which is given by Equation (A.28):

$$R = F(x) = 1 - \exp\left[-\left(\frac{x}{\alpha} \right)^{\beta} \right] \tag{B.20}$$

Equivalently,

$$X = \alpha[-\ln(1 - R)]^{1/\beta} \tag{B.21}$$

Since $(1 - R)$ distributes uniformly in the same way as R in the interval $[0, 1]$, Equation (B.21) becomes

$$X = \alpha(-\ln R)^{1/\beta} \qquad\qquad (B.22)$$

The algorithm for generating the Weibull distribution random variate is as follows:

Step 1: Generate a uniform distribution random number sequence R between $[0, 1]$.

Step 2: Calculate the Weibull distribution random variate X using Equation (B.22).

APPENDIX C

POWER-FLOW MODELS

C.1 AC POWER-FLOW MODELS

C.1.1 Power-Flow Equations

The power-flow equations in the polar coordinate form are as follows:

$$P_i = V_i \sum_{i=1}^{n} V_j (G_{ij} \cos \delta_{ij} + B_{ij} \sin \delta_{ij}) \qquad (i = 1, \ldots, n) \qquad (\text{C.1})$$

$$Q_i = V_i \sum_{i=1}^{n} V_j (G_{ij} \sin \delta_{ij} - B_{ij} \cos \delta_{ij}) \qquad (i = 1, \ldots, n) \qquad (\text{C.2})$$

where P_i and Q_i are the real and reactive bus power injections at Bus i; V_i and δ_i are the magnitude and angle of the voltage at Bus i; $\delta_{ij} = \delta_i - \delta_j$; G_{ij} and B_{ij} are, respectively, the real and imaginary parts of the element in the bus admittance matrix; and n is the number of system buses.

Each bus has four variables (P_i, Q_i, V_i, and δ_i). In order to solve the $4n$-dimensional equations given in Equations (C.1) and (C.2), two of the four variables for each bus have to be prespecified. Generally, P_i and Q_i at load buses are known and they are called PQ buses. P_i and V_i at generator buses are specified and they are called PV buses. V_i and δ_i of one bus in the system have to be specified to adjust the power balance of the system and it is called the swing bus.

Risk Assessment of Power Systems. By Wenyuan Li
ISBN 0-471-63168-X © 2005 the Institute of Electrical and Electronics Engineers, Inc.

C.1.2 Newton–Raphson Method

The Newton–Raphson method is the well-known approach to solve a set of nonlinear simultaneous equations. Equations (C.1) and (C.2) are linearized to yield the following matrix equation:

$$\begin{bmatrix} \Delta P \\ \Delta Q \end{bmatrix} = \begin{bmatrix} H & N \\ J & L \end{bmatrix} \begin{bmatrix} \Delta \delta \\ \Delta V/V \end{bmatrix} \tag{C.3}$$

The Jocobian matrix is a $(n + m - 1)$-dimensional square matrix where n and m are the numbers of all buses and load buses, respectively. $\Delta V/V$ signifies that its elements are $\Delta V_i/V_i$. The elements of the Jocobian matrix are calculated by

$$H_{ij} = \frac{\partial P_i}{\partial \delta_j} = V_i V_j (G_{ij} \sin \delta_{ij} - B_{ij} \cos \delta_{ij}) \tag{C.4}$$

$$H_{ii} = \frac{\partial P_i}{\partial \delta_i} = -Q_i - B_{ii} V_i^2 \tag{C.5}$$

$$N_{ij} = \frac{\partial P_i}{\partial V_j} V_j = V_i V_j (G_{ij} \cos \delta_{ij} + B_{ij} \sin \delta_{ij}) \tag{C.6}$$

$$N_{ii} = \frac{\partial P_i}{\partial V_i} V_i = P_i + G_{ii} V_i^2 \tag{C.7}$$

$$J_{ij} = \frac{\partial Q_i}{\partial \delta_j} = -N_{ij} \tag{C.8}$$

$$J_{ii} = \frac{\partial Q_i}{\partial \delta_i} = P_i - G_{ij} V_i^2 \tag{C.9}$$

$$L_{ij} = \frac{\partial Q_i}{\partial V_j} V_j = H_{ij} \tag{C.10}$$

$$L_{ii} = \frac{\partial Q_i}{\partial V_i} V_i = Q_i - B_{ii} V_i^2 \tag{C.11}$$

The resolution is an iterative process. Given the initial values of V_i and δ_i, the Jocobian matrix is built and Equation (C.3) is solved to find $\Delta \delta_i$ and ΔV_i. Then V_i and δ_i are updated. The process is repeated until a convergence criterion is reached.

C.1.3 Fast Decoupled Method

In power systems, particularly in high-voltage transmission systems, the branch reactance is normally much larger than the branch resistance and the angle differences between two buses are very small. This results in the fact that the element values in the matrix blocks N and J are much smaller than those of H and L. Equation (C.3) can be decoupled by assuming $N = 0$ and $J = 0$. Considering $|(G_{ij} \sin \delta_{ij}| \ll |B_{ij} \cos \delta_{ij}|$ and $|Q_i| \ll |B_{ii}V_i^2|$, the decoupled equation can be further simplified to

$$[\Delta P/V] = [B'][V\Delta\delta] \tag{C.12}$$

$$[\Delta Q/V] = [B''][\Delta V] \tag{C.13}$$

where $[\Delta P/V]$ and $[\Delta Q/V]$ are the vectors whose elements are $\Delta P_i/V_i$ and $\Delta Q_i/V_i$, respectively. $[V\Delta\delta]$ is the vector whose elements are $V_i\Delta\delta_i$. The elements of the constant matrices $[B']$ and $[B'']$ are calculated by

$$B'_{ij} = \frac{-1}{x_{ij}} \tag{C.14}$$

$$B'_{ii} = -\sum_{j \in R_i} B'_{ij} \tag{C.15}$$

$$B''_{ij} = \frac{-x_{ij}}{r_{ij}^2 + x_{ij}^2} \tag{C.16}$$

$$B''_{ii} = -2b_{i0} - \sum_{j \in R_i} B''_{ij} \tag{C.17}$$

where r_{ij} and x_{ij} are the branch resistance and reactance, respectively; b_{i0} is the susceptance between bus i and the ground; and R_i is the set of the buses directly connected to bus i.

C.2 DC POWER-FLOW MODELS

All the notations in this section are the same as defined in Section C.1 except ones specifically defined.

C.2.1 Basic Equation

The DC power-flow equations are based on the following four assumptions:

1. The branch resistance is much smaller than the reactance so that the branch susceptance can be approximated by

$$b_{ij} \approx \frac{-1}{x_{ij}} \tag{C.18}$$

2. The voltage angle difference between the two buses of a branch is small, therefore

$$\sin \delta_{ij} \approx \delta_i - \delta_j$$
$$\cos \delta_{ij} \approx 1.0$$

(C.19)

3. The susceptance between a bus and the ground can be neglected:

$$b_{i0} = b_{j0} \approx 0$$

(C.20)

4. All bus voltage magnitudes are assumed to be 1.0 p.u.

With the above four assumptions, the real power flow in a branch can be calculated by

$$P_{ij} = \frac{\delta_i - \delta_j}{x_{ij}}$$

(C.21)

and, therefore, bus real power injections are

$$P_i = \sum_{j \in R_i} P_{ij} = B'_{ii} \delta_i + \sum_{j \in R_i} B'_{ij} \delta_j \qquad (i = 1, \ldots, n)$$

(C.22)

where B'_{ij} and B'_{ii} are given, respectively, by Equations (C.14) and (C.15).

Using a matrix form, Equation (C.22) can be expressed as:

$$[P] = [B'][\delta]$$

(C.23)

Obviously, this is a set of simple linear algebraic equations and its resolution does not need any iterations. By assuming that bus n is the swing bus and letting $\delta_n = 0$, $[B']$ is a $(n - 1)$-dimensional square matrix. It is exactly the same as the $[B']$ in Equation (C.12).

C.2.2   Line-Flow Equation [84]

By combining Equation (C.23) with (C.21), the following linear relationship between real power injections at buses and line flows can be obtained:

$$[T_p] = [A][P]$$

(C.24)

where T_p is the line flow vector and its elements are the line flows $\{P_{ij}\}$. Matrix $[A]$ is the relationship matrix between power injections and line flows and its dimension is $L \times (n - 1)$, where L is the number of lines and n the number of buses.

Matrix [*A*] can be calculated directly from [*B′*]. Assume that the two buses of line *k* are numbered by *i* and *j*. For *k* = 1, . . . , *L*, the *k*th row of Matrix [*A*] is the solution of the following set of linear equations:

$$[B'][X] = [C] \tag{C.25}$$

where

$$C = \left[0, \ldots, 0, \frac{1}{x_{ij}}, 0, \ldots, 0, -\frac{1}{x_{ij}}, 0, \ldots, 0 \right]^T$$

↑ ith element   ↑ jth element

APPENDIX D

OPTIMIZATION ALGORITHMS

D.1 SIMPLEX METHODS FOR LINEAR PROGRAMMING [85]

The linear programming problem is to minimize (or maximize) a linear objective function while satisfying a set of linear equality and inequality constraints. It has the following standard form:

$$\left.\begin{array}{ll} \min & c^T x \\ \text{subject to} & \\ & Ax = b \\ & 0 \leq x \leq h \end{array}\right\} \tag{D.1}$$

where c, h, and x are the n-dimensional column vectors, b is an m-dimensional column vector, and A an $m \times n$ dimensional matrix.

If there is an inequality constraint of a linear function of x, a nonnegative slack variable can be introduced to convert it to a linear equality constraint. For instance, the inequality $d^T x \geq 0$ can be converted to $d^T x + y = 0$ with $y \geq 0$. Therefore, only the simplex methods for the standard form above are discussed.

D.1.1 Primal Simplex Method

The primal simplex method for a linear programming problem includes the following steps:

Step 1: Determine an initial basic feasible solution using the artificial variable technique and create an initial simplex table:

Risk Assessment of Power Systems. By Wenyuan Li
ISBN 0-471-63168-X © 2005 the Institute of Electrical and Electronics Engineers, Inc.

x	x_1		.	. . .	x_m	x_{m+1}	. . .	x_n	b
	c_1		.	. . .	c_m	c_{m+1}	. . .	c_n	
x_1	1	0		. . .	0	$y_{1,m+1}$	. . .	y_{1n}	y_{10}
x_2	0	1		. . .	0	$y_{2,m+1}$	. . .	y_{2n}	y_{20}
.	.	.	.	. . .	.	.	. . .	.	.
.	.	.	.	. . .	.	.	. . .	.	.
x_i	0	0	. 1 .	0	$y_{i,m+1}$	. . .	y_{in}	y_{i0}	
.	.	.	. . .	.	.	. . .	.	.	
.	.	.	. . .	.	.	. . .	.	.	
x_m	0	0	. 0 .	1	$y_{m,m+1}$	. . .	y_{mn}	y_{m0}	
	0	0	. 0 .	0	r_{m+1}	. . .	r_n	$-z_0$	
	e_1	e_2	. 0 .	e_m	e_{m+1}	. . .	e_n		

In the table, y_{ij} and y_{i0} are the coefficients corresponding to matrix A and vector b, respectively, following each Gaussian elimination step; $r_j = c_j - z_j$ ($j = m + 1, \ldots, n$), where c_j are the coefficients in the objective function of the original problem, which are known as direct cost coefficients; $z_j = \Sigma \, c_i y_{ij}$ are known as composite cost coefficients and r_j are known as relative cost coefficients; $z_0 = \Sigma c_i y_{i0}$ is the value of the objective function at the present step; $e_j = +$ or $-$ ($j = 1, \ldots, n$), which is called the sign row. For the initial basic feasible solution,

$$x_i = \begin{cases} y_{i0} & \text{if } x_i \text{ is a basic variable} \\ 0 & \text{if } x_i \text{ is a nonbasic variable} \end{cases} \tag{D.2}$$

and all the e_j take $+$ signs.

Step 2: Select $r_k = \min \{r_j < 0, j = m + 1, \ldots, n\}$. The kth column is called the pivotal column. If there is no negative r_j, the present solution is already an optimal and feasible solution and the simplex process ends. The values of the variables are determined according to the signs of e_j. If $e_j = +$, $x_j = x_j$, and if $e_j = -$, $x_j = h_j - x_j$. If there still exists negative r_j, go to Step3.

Step 3: Calculate the following three values for the elements in the selected pivotal column:
- h_k
- $\theta_1 = \min\{y_{i0}/y_{ik}\}$ for all $y_{ik} > 0$ (if there is no positive y_{ik}, $\theta_1 = \infty$)
- $\theta_2 = \min\{(y_{i0} - h_i)/y_{ik}\}$ for all $y_{ik} < 0$ (if there is no negative y_{ik}, $\theta_2 = \infty$)

Step 4: Modify the simplex table according to the magnitude of the three values in Step 3:
- If h_k is the minimum, the last column is subtracted by the column that is the product of the kth column and h_k and then the kth column is

multiplied by -1 (including the change of the sign for e_k.). The base remains unchanged.

- If θ_1 is the minimum and θ_1 appears in the qth row, then y_{qk} is selected as the pivot.
- If θ_2 is the minimum and θ_2 appears in the qth row, then $y_{q0(new)} = y_{q0(old)} - h_q$, y_{qq} is multiplied by -1 and the sign of e_q is changed; y_{qk} is selected as the pivot.

Step 5: With the selected pivot element y_{qk}, conduct the Gaussian elimination in the simplex table so that the pivot becomes 1 and the other elements in the pivotal column become 0. The updated simplex table is obtained; go to Step 2.

D.1.2 Dual Simplex Method

The primal simplex method starts from an initial feasible solution and an optimal solution is gradually obtained while retaining feasibility. On the contrary, the dual simplex method begins with an initial basic solution satisfying optimality of the objective function but not meeting feasibility. Feasibility is gradually obtained under the condition that optimality is retained. The dual simplex method can be summarized as follows:

Step 1: Create the initial simplex table of the primal problem corresponding to a dual basic feasible solution x_B, that is, in the simplex table, $r_j \geq 0$ for $j = m + 1, \ldots, n$.

Step 2: If $x_B \geq 0$, that is, there is no negative element in the column b of the simplex table, then an optimal and feasible solution is already reached. If there are any negative elements in the column of b, go to Step 3.

Step 3: Select the smallest value in the negative elements of x_B:

$$\min\{(x_B)_i|(x_B)_i < 0\} = x_q \tag{D.3}$$

The x_q is the leaving base variable. This means that the qth row is the pivotal row.

Step 4: Check all elements of the pivotal row y_{qj} ($j = 1, \ldots, n$). If all the elements $y_{qj} \geq 0$, there is no feasible solution. If there are negative elements in the pivotal row, then

$$\theta = \min\{(z_j - c_j)/y_{qj}|y_{qj} < 0\} = (z_k - c_k)/y_{qk} \tag{D.4}$$

where c_j, z_j, and y_{qj} are the same as defined in the primal simplex method, and x_k is the entering base variable, which means that the kth column is the pivotal column.

Step 5: With the pivot element y_{qk}, conduct the Gaussian elimination to update the simplex table. An updated optimal base B is obtained and then a new dual basic feasible solution is calculated: $x_B = B^{-1}b$. Go to Step 2.

D.2   INTERIOR POINT METHOD FOR NONLINEAR PROGRAMMING [78]

This section describes the primal–dual interior point method that is suitable for nonlinear programming (NLP). The NLP problem has the following mathematical form:

$$\left.\begin{aligned}
\min &\quad f(x) \\
\text{subject to} &\quad g(x) = 0 \\
&\quad \underline{h} \le h(x) \le \overline{h}
\end{aligned}\right\} \tag{D.5}$$

where f is a scalar function, x is a decision variable vector, and g and h correspond to the two vector equations, respectively, with a set of equality and inequality constraints. Note that the inequality $\underline{h} \le h(x) \le \overline{h}$ has a general implication and also includes the inequality constraint of the decision variable vector itself, that is, $\underline{x} \le x \le \overline{x}$.

D.2.1   Optimality and Feasibility Conditions

The inequality constraints in Equations (D.5) can be converted into the equalities by adding nonnegative slack vectors y and z as follows:

$$\left.\begin{aligned}
\min &\quad f(x) \\
\text{subject to} &\\
&\quad g(x) = 0 \\
&\quad h(x) - y - \underline{h} = 0 \\
&\quad -h(x) - z + \overline{h} = 0 \\
&\quad y \ge 0, z \ge 0
\end{aligned}\right\} \tag{D.6}$$

The nonnegative condition on the slack vectors y and z can be guaranteed by introducing them into the logarithmic barrier terms in the objective function so that Equation (D.6) become

$$\left.\begin{aligned}
\min &\quad f(x) - \mu^k \sum_{i=1}^{m} (\ln y_i + \ln z_i) \\
\text{subject to} &\\
&\quad g(x) = 0 \\
&\quad h(x) - y - \underline{h} = 0 \\
&\quad -h(x) - z + \overline{h} = 0
\end{aligned}\right\} \tag{D.7}$$

In Equations (D.7), the logarithmic terms impose the strictly positive conditions on the slack variables so that it is not necessary to explicitly express their nonnegativity constraints. μ^k is called a barrier parameter, where k denotes an iteration index in the resolution process that will be discussed in the following. The equality-constrained optimization equations (D.7) can be solved using the Lagrange multiplier approach. The Lagrangian function $L_\mu(w)$ is constructed as follows:

$$L_{\mu}(w) = f(x) - \mu^k \sum_{i=1}^{m}(\ln y_i + \ln z_i) - \lambda^T g(x) - \gamma^T(h(x) - y - \underline{h}) - \pi^T(-h(x) - z + \overline{h}) \quad \text{(D.8)}$$

where $w = \{x, y, z, \lambda, \gamma, \pi\}$; λ, γ, and π are called dual variables and x, y, and z are called primal variables; and x and λ are the n-dimensional vectors and the other variables are the m-dimensional vectors.

According to the Kuhn–Tucker optimality condition, the local minimum point of the Lagrangian function is reached when its gradient equals zero:

$$\frac{\partial L_{\mu}(w)}{\partial w} = \begin{bmatrix} \left[\dfrac{\partial f(x)}{\partial x}\right] - \left[\dfrac{\partial g(x)}{\partial x}\right]^T \lambda - \left[\dfrac{\partial h(x)}{\partial x}\right]^T \gamma + \left[\dfrac{\partial h(x)}{\partial x}\right]^T \pi \\ \gamma - \mu^k Y^{-1} u \\ \pi - \mu^k Z^{-1} u \\ -g(x) \\ -h(x) + y + \underline{h} \\ h(x) + z - \overline{h} \end{bmatrix} = [0] \quad \text{(D.9)}$$

where $Y = \mathrm{diag}(y_1, y_2, \ldots, y_m)$, $Z = \mathrm{diag}(z_1, z_2, \ldots, z_m)$, and $u = (1, 1, \ldots, 1)^T$. By left-multiplying the second and third term in (D.9) by Y and Z, respectively, we obtain

$$\frac{\partial L_{\mu}(w)}{\partial w} = \begin{bmatrix} \left[\dfrac{\partial f(x)}{\partial x}\right] - \left[\dfrac{\partial g(x)}{\partial x}\right]^T \lambda - \left[\dfrac{\partial h(x)}{\partial x}\right]^T \gamma + \left[\dfrac{\partial h(x)}{\partial x}\right]^T \pi \\ Y\gamma - \mu^k u \\ Z\pi - \mu^k u \\ -g(x) \\ -h(x) + y + \underline{h} \\ h(x) + z - \overline{h} \end{bmatrix} = [0] \quad \text{(D.10)}$$

In Equation (D.10), the first term, along with $\gamma \geq 0$ and $\pi \geq 0$, assures the dual feasibility; the fourth, fifth, and sixth terms, together with $y \geq 0$ and $z \geq 0$, assure the primal feasibility; the second and third terms are called the complementarity conditions.

The primal–dual interior point method for solving Equations (D.5) is an iteration process. Given an initial value of μ^0 and a starting point w^0, the set of nonlinear equations in Equation (D.10) is solved, a step length in the correction direction is calculated, and then the variable vector w is updated. The process is repeated with the reduced barrier parameter μ^k. During the iteration, the nonnegativity of the slack variables and multipliers must be satisfied at every point. The iteration terminates when the primal unfeasibility, dual unfeasibility, and complementarity gap are smaller than prespecified tolerances. The resolution process gradually reaches optimality and feasibility as μ^k approaches zero.

D.2.2 Procedure for the Algorithm

The primal–dual interior point method includes the following steps:

Step 1: Specify an initial value of μ^0 and select the starting point w^0 that strictly meets the positive conditions.

Step 2: Solve the correction equations of Equation (D.10) at the current point to obtain a correction direction. By applying the Newton–Raphson method, the following correction equations can be built up:

$$
\begin{bmatrix}
\left[\dfrac{\partial^2 L_\mu}{\partial x^2}\right] & 0 & 0 & -\left[\dfrac{\partial g}{\partial x}\right]^T & -\left[\dfrac{\partial h}{\partial x}\right]^T & \left[\dfrac{\partial h}{\partial x}\right]^T \\
0 & \Gamma & 0 & 0 & Y & 0 \\
0 & 0 & \Pi & 0 & 0 & Z \\
-\left[\dfrac{\partial g}{\partial x}\right] & 0 & 0 & 0 & 0 & 0 \\
-\left[\dfrac{\partial h}{\partial x}\right] & I & 0 & 0 & 0 & 0 \\
\left[\dfrac{\partial h}{\partial x}\right] & 0 & I & 0 & 0 & 0
\end{bmatrix}
\begin{bmatrix} \Delta x \\ \Delta y \\ \Delta z \\ \Delta \lambda \\ \Delta \gamma \\ \Delta \pi \end{bmatrix}
=
\begin{bmatrix} b_x \\ b_y \\ b_z \\ b_\lambda \\ b_\gamma \\ b_\pi \end{bmatrix}
\qquad \text{(D.11)}
$$

where

$$\Gamma = \mathrm{diag}(\gamma_1,\ \gamma,\ \ldots,\ \gamma_m),\ \Pi = \mathrm{diag}(\pi_1,\ \pi,\ \ldots,\ \pi_m),\ \text{and } I = \mathrm{diag}(1,\ 1,\ \ldots,\ 1)$$

$$
\left[\frac{\partial^2 L_\mu}{\partial x^2}\right] = \left[\frac{\partial f^2(x)}{\partial x^2}\right] - \left[\frac{\partial g^2(x)}{\partial x^2}\right]^T \lambda - \left[\frac{\partial h^2(x)}{\partial x^2}\right]^T \gamma + \left[\frac{\partial h^2(x)}{\partial x^2}\right]^T \pi \quad \text{(D.12)}
$$

$$
\begin{bmatrix} b_x \\ b_y \\ b_z \\ b_\lambda \\ b_\gamma \\ b_\pi \end{bmatrix}
=
\begin{bmatrix}
-\left[\dfrac{\partial f(x)}{\partial x}\right] + \left[\dfrac{\partial g(x)}{\partial x}\right]^T \lambda + \left[\dfrac{\partial h(x)}{\partial x}\right]^T \gamma - \left[\dfrac{\partial h(x)}{\partial x}\right]^T \pi \\
-Y\gamma + \mu^k u \\
-Z\pi + \mu^k u \\
g(x) \\
h(x) - y - \underline{h} \\
-h(x) - z + \overline{h}
\end{bmatrix}
\qquad \text{(D.13)}
$$

It should be noted that the coefficient matrix and the right-side vector in Equation (D.11) are the calculated values at the current point in the iteration. The superscript for the iteration has been omitted for simplicity.

Step 3: Update the primal and dual variables in the correction direction using the following formulas:

$$
\begin{bmatrix} x \\ y \\ z \end{bmatrix}^{(k+1)} = \begin{bmatrix} x \\ y \\ z \end{bmatrix}^{(k)} + \eta\alpha_p^k \begin{bmatrix} \Delta x \\ \Delta y \\ \Delta z \end{bmatrix}^{(k)}
\tag{D.14}
$$

$$
\begin{bmatrix} \lambda \\ \gamma \\ \pi \end{bmatrix}^{(k+1)} = \begin{bmatrix} \lambda \\ \gamma \\ \pi \end{bmatrix}^{(k)} + \eta\alpha_d^k \begin{bmatrix} \Delta\lambda \\ \Delta\gamma \\ \Delta\pi \end{bmatrix}^{(k)}
\tag{D.15}
$$

where η is a scalar between $(0,1)$ that ensures the positive conditions in the next point and whose value is often given as 0.9995; α_p^k and α_d^k are the step lengths for the primal and dual variables respectively and they are selected by

$$
\alpha_p^k = \min\left\{ 1, \min_{\Delta y_i \le -\delta}\left\{ \frac{y_i}{|\Delta y_i|} \right\}, \min_{\Delta z_i \le -\delta}\left\{ \frac{z_i}{|\Delta z_i|} \right\} \right\}
\tag{D.16}
$$

$$
\alpha_d^k = \min\left\{ 1, \min_{\Delta\gamma_i \le -\delta}\left\{ \frac{\gamma_i}{|\Delta\gamma_i|} \right\}, \min_{\Delta\pi_i \le -\delta}\left\{ \frac{\pi_i}{|\Delta\pi_i|} \right\} \right\}
\tag{D.17}
$$

where δ is a given tolerance.

Step 4: Check that the following convergence criteria are met: the barrier parameter μ^k is sufficiently small, the equality constraints are satisfied, and the changes in the objective function and variables between two iterations are negligible. The criteria are mathematically expressed as:

$$
\left.\begin{aligned} \mu^k &\le \varepsilon_0 \\ \|\mathbf{g}(\mathbf{x})\| &\le \varepsilon_1 \\ \|\Delta\mathbf{x}\| &\le \varepsilon_2 \\ \frac{|f(\mathbf{x}^k) - f(\mathbf{x}^{k-1})|}{|f(\mathbf{x}^k)|} &\le \varepsilon_3 \end{aligned}\right\}
\tag{D.18}
$$

If the convergence criteria are satisfied at the new point, the iteration ends. Otherwise, go to Step 5.

Step 5: Update the barrier parameter using Equation (D.19) and return to Step 2.

$$
\mu^{k+1} = \tau^k \frac{(\mathbf{y}^k)^T \gamma^k + (\mathbf{z}^k)^T \pi^k}{2m}
\tag{D.19}
$$

where τ^k is a positive parameter smaller than 1.0. It is often chosen as $\tau^k = \max\{0.99\,\tau^{k-1}, 0.1\}$, with $\tau^0 = 0.2 \sim 0.3$. The m is the dimension of $\mathbf{y}$ or $\mathbf{z}$.

APPENDIX E

THREE PROBABILITY
DISTRIBUTION TABLES

The three probability distribution tables that have been used for the parameter interval estimation in Chapter 3 are provided in this Appendix.

TABLE 1 Relationship between area Q and z under the standard normal distribution

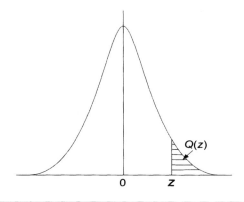

					Q					
z	0.00	0.01	0.02	0.03	0.04	0.05	0.06	0.07	0.08	0.09
0.0	0.5000	0.4960	0.4920	0.4880	0.4840	0.4801	0.4761	0.4721	0.4681	0.4641
0.1	0.4602	0.4562	0.4522	0.4483	0.4443	0.4404	0.4364	0.4325	0.4286	0.4247
0.2	0.4207	0.4168	0.4129	0.4090	0.4052	0.4013	0.3974	0.3936	0.3897	0.3859
0.3	0.3821	0.3783	0.3745	0.3707	0.3669	0.3632	0.3594	0.3557	0.3520	0.3483
0.4	0.3446	0.3409	0.3372	0.3336	0.3300	0.3264	0.3228	0.3192	0.3156	0.3121
0.5	0.3085	0.3050	0.3015	0.2981	0.2946	0.2912	0.2877	0.2843	0.2810	0.2776
0.6	0.2743	0.2709	0.2676	0.2643	0.2611	0.2578	0.2546	0.2514	0.2483	0.2451
0.7	0.2420	0.2389	0.2358	0.2327	0.2296	0.2266	0.2236	0.2206	0.2177	0.2148
0.8	0.2119	0.2090	0.2061	0.2033	0.2005	0.1977	0.1949	0.1922	0.1894	0.1867
0.9	0.1841	0.1814	0.1788	0.1762	0.1736	0.1711	0.1685	0.1660	0.1635	0.1611
2.0	0.1587	0.1562	0.1539	0.1515	0.1492	0.1469	0.1446	0.1423	0.1401	0.1379
1.1	0.1357	0.1335	0.1314	0.1292	0.1271	0.1251	0.1230	0.1210	0.1190	0.1170
1.2	0.1151	0.1131	0.1112	0.1093	0.1075	0.1056	0.1038	0.1020	0.1003	0.0985
1.3	0.0968	0.0951	0.0934	0.0918	0.0901	0.0885	0.0869	0.0853	0.0838	0.0823
1.4	0.0808	0.0793	0.0778	0.0764	0.0749	0.0735	0.0721	0.0708	0.0694	0.0681
1.5	0.0668	0.0655	0.0643	0.0630	0.0618	0.0606	0.0594	0.0582	0.0571	0.0559
1.6	0.0548	0.0537	0.0526	0.0516	0.0505	0.0495	0.0485	0.0475	0.0465	0.0455
1.7	0.0446	0.0436	0.0427	0.0418	0.0409	0.0401	0.0392	0.0384	0.0375	0.0367
1.8	0.0359	0.0351	0.0344	0.0336	0.0329	0.0322	0.0314	0.0307	0.0301	0.0294
1.9	0.0287	0.0281	0.0274	0.0268	0.0262	0.0256	0.0250	0.0244	0.0239	0.0233
2.0	0.0228	0.0222	0.0217	0.0212	0.0207	0.0202	0.0197	0.0192	0.0188	0.0183
2.1	0.0179	0.0174	0.0170	0.0166	0.0162	0.0158	0.0154	0.0150	0.0146	0.0143
2.2	0.0139	0.0136	0.0132	0.0129	0.0125	0.0122	0.0119	0.0116	0.0113	0.0110
2.3	0.0107	0.0104	0.0102	0.0099	0.0096	0.0094	0.0091	0.0089	0.0087	0.0084
2.4	0.0082	0.0080	0.0078	0.0075	0.0073	0.0071	0.0069	0.0068	0.0066	0.0064
2.5	0.0062	0.0060	0.0059	0.0057	0.0055	0.0054	0.0052	0.0051	0.0049	0.0048
2.6	0.0047	0.0045	0.0044	0.0043	0.0041	0.0040	0.0039	0.0038	0.0037	0.0036
2.7	0.0035	0.0034	0.0033	0.0032	0.0031	0.0030	0.0029	0.0028	0.0027	0.0026
2.8	0.0026	0.0025	0.0024	0.0023	0.0023	0.0022	0.0021	0.0021	0.0020	0.0019
2.9	0.0019	0.0018	0.0018	0.0017	0.0016	0.0016	0.0015	0.0015	0.0014	0.0014
3.0	0.0013	0.0013	0.0013	0.0012	0.0012	0.0011	0.0011	0.0011	0.0010	0.0010

TABLE 2 Relationship between area α and $t_\alpha(n)$ under the t-distribution

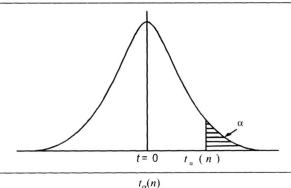

$t = 0 \qquad t_\alpha(n)$

			$t_\alpha(n)$		
n	$\alpha = 0.10$	$\alpha = 0.05$	$\alpha = 0.025$	$\alpha = 0.01$	$\alpha = 0.005$
1	3.078	6.314	12.706	31.821	63.657
2	1.886	2.920	4.303	6.965	9.925
3	1.638	2.353	3.182	4.541	5.841
4	1.533	2.132	2.776	3.747	4.604
5	1.476	2.015	2.571	3.365	4.032
6	1.440	1.943	2.447	3.143	3.707
7	1.415	1.895	2.365	2.998	3.499
8	1.397	1.860	2.306	2.896	3.355
9	1.383	1.833	2.262	2.821	3.250
10	1.372	1.812	2.228	2.764	3.169
11	1.363	1.796	2.201	2.718	3.106
12	1.356	1.782	2.179	2.681	3.055
13	1.350	1.771	2.160	2.650	3.012
14	1.345	1.761	2.145	2.624	2.977
15	1.341	1.753	2.131	2.602	2.947
16	1.337	1.746	2.120	2.583	2.921
17	1.333	1.740	2.110	2.567	2.898
18	1.330	1.734	2.101	2.552	2.878
19	1.328	1.729	2.093	2.539	2.861
20	1.325	1.725	2.086	2.528	2.845
21	1.323	1.721	2.080	2.518	2.831
22	1.321	1.717	2.074	2.508	2.819
23	1.319	1.714	2.069	2.500	2.807
24	1.318	1.711	2.064	2.492	2.797
25	1.316	1.708	2.060	2.485	2.787
26	1.315	1.706	2.056	2.479	2.779
27	1.314	1.703	2.052	2.473	2.771
28	1.313	1.701	2.048	2.467	2.763
29	1.311	1.699	2.045	2.462	2.756
500	1.283	1.648	1.965	2.334	2.586
1000	1.282	1.646	1.962	2.330	2.581

TABLE 3 Relationship between area α and $\chi_\alpha^2(n)$ under the χ^2 distribution

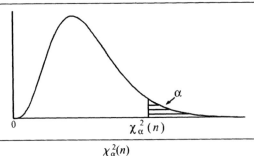

$$\chi_\alpha^2(n)$$

n	$\alpha = 0.995$	$\alpha = 0.99$	$\alpha = 0.975$	$\alpha = 0.95$	$\alpha = 0.05$	$\alpha = 0.025$	$\alpha = 0.01$	$\alpha = 0.005$
1	0.0000393	0.000157	0.000982	0.00393	3.841	5.024	6.635	7.879
2	0.0100	0.0201	0.0506	0.103	5.991	7.378	9.210	10.597
3	0.0717	0.115	0.216	0.352	7.815	9.348	11.345	12.838
4	0.207	0.297	0.484	0.711	9.488	11.143	13.277	14.860
5	0.412	0.554	0.831	1.145	11.070	12.832	15.086	16.750
6	0.676	0.872	1.237	1.635	12.592	14.449	16.812	18.548
7	0.989	1.239	1.690	2.167	14.067	16.013	18.475	20.278
8	1.344	1.646	2.180	2.733	15.507	17.535	20.090	21.955
9	1.735	2.088	2.700	3.325	16.919	19.023	21.666	23.589
10	2.156	2.558	3.247	3.940	18.307	20.483	23.209	25.188
11	2.603	3.053	3.816	4.575	19.675	21.920	24.725	26.757
12	3.074	3.571	4.404	5.226	21.026	23.337	26.217	28.300
13	3.565	4.107	5.009	5.892	22.362	24.736	27.688	29.819
14	4.075	4.660	5.629	6.571	23.685	26.119	29.141	31.319
15	4.601	5.229	6.262	7.261	24.996	27.488	30.578	32.801
16	5.142	5.812	6.908	7.962	26.296	28.845	32.000	34.267
17	5.697	6.408	7.564	8.672	27.587	30.191	33.409	35.718
18	6.265	7.015	8.231	9.390	28.869	31.526	34.805	37.156
19	6.844	7.633	8.907	10.117	30.144	32.852	36.191	38.582
20	7.434	8.260	9.591	10.851	31.410	34.170	37.566	39.997
21	8.034	8.897	10.283	11.591	32.671	35.479	38.932	41.401
22	8.643	9.542	10.982	12.338	33.924	36.781	40.289	42.796
23	9.260	10.196	11.689	13.091	35.172	38.076	41.638	44.181
24	9.886	10.856	12.401	13.848	36.415	39.364	42.980	45.558
25	10.520	11.524	13.120	14.611	37.652	40.646	44.314	46.928
26	11.160	12.198	13.844	15.379	38.885	41.923	45.642	48.290
27	11.808	12.879	14.573	16.151	40.113	43.194	46.963	49.645
28	12.461	13.565	15.308	16.928	41.337	44.461	48.278	50.993
29	13.121	14.256	16.047	17.708	42.557	45.722	49.588	52.336
30	13.787	14.953	16.791	18.493	43.773	46.979	50.892	53.672
40	20.707	22.164	24.433	26.509	55.758	59.342	63.691	66.766
50	27.991	29.707	32.357	34.764	67.505	71.420	76.154	79.490
60	35.534	37.485	40.482	43.188	79.082	83.298	88.379	91.952
70	43.275	45.442	48.758	51.739	90.531	95.023	100.425	104.215
80	51.172	53.540	57.153	60.391	101.879	106.629	112.329	116.321
90	59.196	61.754	65.647	69.126	113.145	118.136	124.116	128.299
100	67.328	70.065	74.222	77.929	124.342	129.561	135.807	140.170

REFERENCES

1. R. Billinton and W. Li, *Reliability Assessment of Electric Power Systems Using Monte Carlo Methods,* Plenum Press, New York and London, 1994.

2. R. Billinton and R. N. Allan, *Reliability Evaluation of Engineering Systems: Concepts and Techniques* (second edition), Plenum Press, New York and London, 1992.

3. R. Billinton and R. N. Allan, *Reliability Evaluation of Power Systems* (second edition), Plenum Press, New York and London, 1996.

4. E. J. Henley and H. Kumamoto, *Probabilistic Risk Assessment,* IEEE Press, New York, 1992.

5. R. Billinton, R. N. Allan, and L. Salvaderi, *Applied Reliability Assessment in Electric Power Systems,* IEEE Press, New York, 1991.

6. J. Endrenyi, *Reliability Modeling in Electric Power Systems,* Wiley, Chichester, UK, 1978.

7. IEEE Gold Book, *Design of Reliable Industrial and Commercial Power Systems,* IEEE Std 493-1997, New York, 1998.

8. IEEE, *Reliability Assessment of Composite Generation and Transmission Systems,* IEEE tutorial course text 90EH0311-1-PWR, 1990.

9. IEEE, *Reliability Data* (corrected edition), IEEE Std 500-1984, New York, 1993.

10. EPRI Report, *Framework for Stochastic Reliability of Bulk Power System,* TR-110048, Palo Alto, California, 1998.

11. CIGRE Working Group 38.03, *Power System Reliability Analysis—Application Guide,* CIGRE Publication, Paris, 1988.

12. R. A. Johnson, *Miller and Freund's Probability and Statistics for Engineers* (fifth edition), Prentice-Hall, Inc. Englewood Cliffs, New Jersey, 1994.

13. Z. Fang, *Computer Simulation and Monte Carlo Method,* Publishing House of Beijing Industrial Institute, 1988.

14. R. Y. Rubinstein, *Simulation and Monte Carlo Method,* Wiley, New York, 1981.

Risk Assessment of Power Systems. By Wenyuan Li
ISBN 0-471-63168-X © 2005 the Institute of Electrical and Electronics Engineers, Inc.

15. N. R. Mann, R. E. Schafer, and N. D. Singpurwalla, *Methods for Statistical Analysis of Reliability and Life Data*, Wiley, New York, 1974.

16. John E. Freund, *Mathematical Statistics*, Prentice-Hall, Inc., Englewood Cliffs, N.J., 1962.

17. C. S. Park, *Contemporary Engineering Economics*, Addison-Wesley, Reading, Massachusetts, 1993.

18. D. Young, *Modern Engineering Economy*, Wiley, New York, 1993.

19. R. N. Allan, R. Billinton, S. M. Shahidehpour, and C. Singh, "Bibliography on the Application of Probability Methods in Power System Reliability Evaluation, 1982–1987," *IEEE Trans. on Power Systems, 3*, 1555–1564, 1988.

20. R. N. Allan R. Billinton, A. M. Briepohl, and C. H. Grigg, "Bibliography on the Application of Probability Methods in Power System Reliability Evaluation, 1987–1991," *IEEE Trans. on Power Systems, PWRS-9*, 1994.

21. W. Li, "Evaluating Mean Life of Power System Equipment with Limited Aging Failure Data," *IEEE Trans. on Power Systems, 19*, 1, 236–242, February 2004.

22. W. Li, S. Pai, M. Kwok, and J. Sun, "Determining Rational Redundancy of 500 kV Reactors in Transmission Systems Using a Probability Based Economic Analysis Approach: BCTC's Practice," *IEEE Trans. on Power Systems, 19*, 1, 325–329, February 2004.

23. W. Li and J. K. Korczynski, "A Reliability Based Approach to Transmission Maintenance Planning and Its Application in BC Hydro System," *IEEE Trans. on Power Delivery, 19*, 1, 303–308, January 2004.

24. W. Li and R Billinton, "Common Cause Outage Models in Power System Reliability Evaluation," *IEEE Trans. on Power Systems, 18*, 2, 966–968, May 2003.

25. W. Li, "Incorporating Aging Failures in Power System Reliability Evaluation," *IEEE Trans. on Power Systems*, 918–923, August 2002.

26. W. Li and J. K. Korczynski, "Risk Evaluation of Transmission Systems Operation Modes: Concept, Method and Application," Paper No. 02WM012, presented at IEEE Winter Meeting, New York, January 27–31, 2002.

27. W. Li and S. Pai, "Evaluating Unavailability of Equipment Aging Failures," *IEEE Power Engineering Review*, 52–54, February, 2002.

28. W. Li, J. K. Korczynski, and G. Armanini, "Application of Probabilistic Reliability Evaluation in Transmission System Operation: BC Hydro's Approach and Experience," in *Proceedings of PMAPS 2002*, Paper No. NE20, Naples, Italy, September 22–26, 2002.

29. E. Vaahedi, W. Li, T. Chia, and H. Dommel, "Large Scale Probabilistic Transient Stability Assessment Using B.C. Hydro's On-line Tools," *IEEE Trans. on Power Systems, 15*, 2, 661–667, May 2000.

30. W. Li, E. Vaahedi, Y. Mansour, "Determining Number and Timing of Substation Spare Transformers Using A Probabilistic Cost Analysis Approach," *IEEE Trans. on Power Delivery, 14*, 3, 934–939, July 1999.

31. W. Li, Y. Mansour, E. Vaahedi, and D. N. Pettet, "Incorporation of Voltage Stability Operating Limits in Composite System Adequacy Assessment: BC Hydro's Experience," *IEEE Trans. on Power Systems, 13*, 4, 1279–1284,November 1998.

32. W. Li and F. P. P. Turner, "Development of Probabilistic Transmission Planning Methodology at BC Hydro," in *Proceedings of PMAPS 1997*, Vancouver, September 21–25, 1997, pp. 25–31, 1997.

33. Y. Mansour and W. Li, "Selecting Location and Size of Co-generations Connected to Area Transmission Systems Based on Impacts on System Reliability," presented at International V SEPOPE Symposium, Recife, Brazil, May 19–24, 1996.

34. W. Li, Y. Mansour, J. K. Korczynski, and B. J. Mills, "Application of Transmission Reliability Assessment in Probabilistic Planning of BC Hydro Vancouver South Metro System," *IEEE Trans. on Power Systems, 10,* 2, 964–970, 1995.

35. W. Li, Y. Mansour, B.J. Mills, and J.K. Korczynski, "Composite System Reliability Evaluation and Probabilistic Planning: BC Hydro's Practice," presented at 1995 Canadian Electric Association Conference, Vancouver, March, 1995.

36. IEEE Task Force, "The IEEE Reliability Test System—1996," presented at 1996 IEEE Winter Meeting, Paper 96WM 326-9 PWRS, Baltimore, MD, January 21–25, 1996.

37. A. F. Vojdani, R. D. Williams, W. Gambel, W. Li, L. Eng, and B. N. Suddeth, "Experience with Application of Reliability and Value of Service Analysis in System Planning," *IEEE Trans. on Power Systems, 11,* 1489–1495, August, 1996.

38. W. Li and R. Billinton, "Incorporation and Effects of Supporting Polices in Multi-Area Generation System Reliability Assessment," *IEEE Trans. on Power Systems, 8,* 3, 1061–1067, 1993.

39. W. Li and R. Billinton, "A Minimum Cost Assessment Method for Composite Generation and Transmission System Expansion Planning," *IEEE Trans. on Power Systems, 8,* 2, 628–635, 1993.

40. R. Billinton and W. Li, "A System State Transition Sampling Method for Composite System Reliability Evaluation," *IEEE Trans. on Power Systems, 8,* 3, 761–770, 1993.

41. R Billinton and W. Li, "A Monte Carlo Method for Multi-Area Generation System Reliability Assessment," *IEEE Trans. on Power Systems, 7,* 4, 1487–1492, 1992.

42. R. Billinton and W. Li, "A Hybrid Approach for Reliability Evaluation of Composite Generation and Transmission Systems Using Monte Carlo Simulation and Enumeration Technique," *IEEE Proceedings C, 138,* 3, 233–241, May 1991.

43. R. Billinton and W. Li, "A Novel Method for Incorporating Weather Effects in Composite System Adequacy Evaluation," *IEEE Trans. on Power Systems, 6,* 3, 1154–1160, 1992.

44. W. Li and R. Billinton, "Effects of Bus Load Uncertainty and Correlation in Composite System Adequacy Evaluation," *IEEE Trans. on Power Systems, 6,* 4, 1522–1529, 1991.

45. R Billinton and W. Li, "Consideration of Multi-State Generating Unit Models in Composite System Adequacy Assessment Using Monte Carlo Simulation," *Canadian Journal of Electrical and Computer Engineering, 17,* 1, 1992.

46. R. Billinton and W. Li, "Direct Incorporation of Load Variations in Monte Carlo Simulation of Composite System Adequacy," presented at Inter-RAMQ Conference for the Electric Power Industry, Philadelphia, PA, Aug. 25–28, 1992.

47. R Billinton and W. Li, "Composite System Reliability Assessment Using a Monte Carlo Approach," presented at The Third International Conference on Probabilistic Methods Applied to Power Systems (PMAPS), London, July 3–5, 1991.

48. R. Billinton and W. Li, "Applications of Monte Carlo Simulation in Power System Adequacy Assessment," *CEA Trans., 30,* 3, 1991.

49. W. Li, "Monte Carlo Reliability Evaluation for Large Scale Composite Generation and Transmission Systems," *Journal of Chongqing University, 12,* 3, 92–98, May 1989.

50. W. Li, "Reliability Assessment of Two Transmission Line Connections Associated with Substations," BC Hydro, RSP-R-002, April 9, 1992.

51. W. Li, "Reliability Assessment of South Metro System Planning Alternatives—Part I," BC Hydro, RSP-R-003, August 10, 1992.

52. W. Li, "Reliability Assessment of South Metro System Planning Alternatives—Part II," BC Hydro, RSP-R-005, September 17, 1992.

53. W. Li, "Probabilistic Planning of IPP in South Vancouver Metro System," BC Hydro, RSP-R-008, March 24, 1993.

54. W. Li, "Reliability Study for Vancouver Island Supply Alternatives," BC Hydro, RSP-R-011, September 17, 1993.

55. W. Li, "Reliability Worth of HVDC Reliability Assurance Project," BC Hydro, RSP-R-015, (Part I: RSP-R-015A, Jan. 17, 1994; Part II: RSP-R-015B, Feb. 4, 1994).

56. W. Li, "Reliability Indices for BC Hydro Composite Generation and 500/230 kV Transmission System," BC Hydro, RSP-R-017, March 8, 1994.

57. W. Li, "Reliability Worth of HVDC Annual Maintenance Scheduling," BC Hydro, RSP-R-018, March 21, 1994.

58. W. Li, "Probability Distribution and Expectation of HVDC Rating," BC Hydro, RSP-R-022, September 19, 1994.

59. W. Li, "Comparison between Two Connection Schemes for IPP Access to MASSET Power Supply System," BC Hydro, RSP-R-024, (Part I: RSP-R-024A, December 16, 1994; Part II: RSP-R-024B, April 25, 1995; Addendum: RSP-R-024C, June 29, 1995).

60. W. Li, "Risk Evaluation for Not Adding Third Transformer at Walters Substation," BC Hydro, STS-940-R5-001, March 23, 1995.

61. W. Li, "Reliability Evaluation of Five Short Listed IPP Proposals in Vancouver Island" BC Hydro, STS-940-R5-005, (Part I: Portalberni, STS-940-R5-005A; Part II: Kokish, STS-940-R5-005B; Part III: Zeballos, STS-940-R5-005C; Part VI: Harmac, STS-940-R5-005D; Part V: Island, STS-940-R5-005E), October 27, 1995.

62. W. Li, "Reliability and Probabilistic Cost Analysis of Mobile Transformer for Single Transformer Substation Group," BC Hydro, STS-940-R5-007, (Part I: Lower Mainland, STS-940-R5-007A, November 24, 1995; Part II: South Interior, STS-940-R5-007B, November 30, 1995; Part III: North Region, STS-940-R5-007C, December 10, 1995).

63. W. Li, "Reliability Assessment and Probabilistic Planning of North Metro System Alternatives," BC Hydro, STS-940-R5-008, Jan. 26, 1996.

64. W. Li, "Mean Time Between Failures Analysis for Macmillan and Bloedel Motor Power Supply at Stations E and H," Project Report for Powertech Labs Inc., Project No. 7986-27, August 23, 1996.

65. W. Li, "Reliability Based Transmission Service Pricing—Review," BC Hydro, CCT-R-002, August 29, 1996.

66. W. Li, "Monthly Reliability Profile of BC Hydro 500 kV System with a Wheeling Customer from REV to CUSTER," BC Hydro, CCT-R-003, December 12, 1996.

67. W. Li, "Reliability Based Transmission Service Pricing—Method and Examples," BC Hydro, CCT-R-004, September 15, 1998.

68. W. Li, "Impacts of Keogh Generation on Upper NVI System Reliability," BC Hydro, CCT-R-005, August 23, 1999.

69. W. Li, "Risk Comparison between Two Planned Outage Schedules for 5L29, 5L30 and 5L31," BC Hydro, CCT-R-006, September 16, 1999.

70. W. Li, "Impacts of New Generators, Wheeling Customers and Network Load Growths on Grid Transmission System Reliability," BC Hydro, CCT-R-008, Jan. 12, 2000.

71. W. Li, "Methods for Determining Unit Interruption Cost," BC Hydro, CCT-R-009, Jan. 19, 2000.

72. W. Li, "Operation Risk Evaluation and Lowest Risk Scheduling of 2L39 Replacement," BC Hydro, CCT-R-010, March 30, 2000.

73. W. Li, "Risk Evaluation of Operation Alternatives in Metro Area with 2L39/2L40 Out-Of-Service," BC Hydro, CCT-R-013, October 5, 2000.

74. W. Li, "Unavailability of 500 kV Reactors and Potential Economic Loss in Export to USA," BC Hydro, CCT-R-014, August 2, 2001.

75. W. Li, "Risk Evaluation of Metro System Operation Modes with New 2L39 In-Service during 2L40 Replacement," BC Hydro, CCT-R-016, January 4, 2002.

76. W. Li, "Failure Event Enumeration Technique for Operation Risk Evaluation of a Simple Transmission System," BC Hydro, CCT-R-017, March 21, 2002.

77. W. Li, "Reliability Based Transmission Pricing," BC Hydro report: Project No. 2bi-SRD P96-052-P0/1486, Strategic Research and Development Program, March 2002.

78. A. V. Fiacco and G. P. McCormick, *Nonlinear Programming: Sequential Unconstrained Minimization Techniques,* Wiley, New York, 1968.

79. Mathematics Manual Editing Group, *Mathematics Manual,* People's Education Press, Beijing, 1997.

80. Canadian Electricity Association, *Generation Equipment Status Annual Report—Equipment Reliability Information Systems,* 1995.

81. Canadian Electricity Association, *Generation Equipment Status Annual Report—Equipment Reliability Information Systems,* 2001.

82. Canadian Electricity Association, *Forced Outage Performance of Transmission Equipment—Equipment Reliability Information Systems,* 2001.

83. EPRI Report, *The Cost of Power Disturbances to Industrial and Digital Economy Companies,* prepared by PRIMEN, June 29, 2001.

84. W. Li, *Secure and Economic Operation of Power Systems—Models and Methods,* Chongqing University Publishing House, 1989.

85. D. G. Luenberger, *Introduction to Linear and Nonlinear Programming,* Addison-Wesley, Reading, Massachusetts, 1973.

86. A. J. Wood and B. F. Wollenberg, *Power Generation, Operation and Control,* Wiley, New York, 1984.

87. J. Wu, Y. Qin, and D. Zhang, *Power Systems,* Power Industry Press, Beijing, 1980.

88. J. Pouget, *Réseaux Électriques,* Électricité de France, Paris, 1974.

INDEX

ABOUT THE AUTHOR

Dr. Wenyuan Li is currently a specialist engineer at British Columbia Transmission Corporation, Vancouver, Canada. He is a Fellow of IEEE and an honorable advisory professor of Chongqing University in China. He obtained his Bachelor degree from Tsinghua University in 1968 and his Master degree and Ph.D. from Chongqing University in 1982 and 1987, all in Electrical Engineering. He was a Post-Doctoral Fellow at the University of Saskatchewan from 1989 to 1991. Before he came to Canada, he worked at the Research and Development Division of EDF in France as a visiting scholar from 1983 to 1985 and Chongqing University as a full professor from 1987 to 1989. Dr. Li has published four books and numerous technical papers in economic operation, optimization, planning, and reliability assessment of power systems. From 1991 to 2004, his work with BC Hydro and British Columbia Transmission Corporation, was dedicated to practical applications of risk evaluation to electric utility systems and has produced more than 50 technical reports based on the actual projects. He developed many computing programs that have been used in electrical utilities for years and have created huge benefits to the utilities. Dr. Li also was committed academic activities such as IEEE PES general meetings, technical subcommittees, and paper reviews. He also serves on the editorial board of the magazine *Electric Power Components and Systems* and served on the technical advisory committee of the international conference "Probabilistic Methods Applied to Power Systems." He delivered tutorials and seminars at different international conferences and industrial workshops and also provided professional development courses to many engineers in the power industry. Dr. Li was the winner of 1996 "Outstanding Engineer Award" of IEEE Canada due to his contributions in power system reliability and probabilistic planning. His biography was included in the second edition of *Five Thousand Personalities of The World* with the citation "Distinguished Professional Achievement."

Risk Assessment of Power Systems. By Wenyuan Li   **325**
ISBN 0-471-63168-X © 2005 the Institute of Electrical and Electronics Engineers, Inc.

Lightning Source UK Ltd.
Milton Keynes UK

177753UK00001B/40/P